Beginning Remote Video Astronomy …

So, what's the matter?

Practical solutions to questions when
starting video assisted Astronomy remotely *

Jim Meadows

* From inside your home or vehicle where it is warm

Preface

This is my third book of practical answers to problems and questions when beginning a new area of Astronomy. My first book, "Beginning Astronomy with a Celestron … So, what's the matter?", dealt with finding solutions to problems when starting Astronomy with a computerized telescope. As I described in that book, after many years away from my college / early marriage hobby, I decided to look into the latest equipment available and get back into Astronomy. In recent years with advances in digital photography, video astronomy, Wi-Fi and other technology, I began to think about how I might even be able to use a telescope outside while staying inside my house.

I started with the Celestron 8se, a GoTo Alt-Az mount. Within a couple of days of receiving it, I had learned how to control it from my laptop using downloaded software, written a simple program on my laptop to move the telescope around and tested it the next day. I then ordered a low-end imaging camera for $100. I quickly realized there were various issues I needed to work through to really learn how to use this new technology.

Later I upgraded to a Celestron Equatorial Mount for improved GoTo accuracy and tracking. This led to my second book, "Beginning Astronomy with a Celestron Equatorial Mount … So, what's the matter?", as I ran into many of the common issues you may encounter when switching to an equatorial mount.

Over the last year I have found out how to accomplish my original objective, and can now sit inside by the fire while controlling my telescope out on the driveway, and I can see night sky wonders using video assisted observing that I could not see through an eyepiece! In this process I found many things that didn't work and some that worked well. I explored a wide range of video from using modified webcams to the excellent MallinCam Extreme X2. I looked into both wired and Wi-Fi methods to control my mounts remotely. Out of this came my third book, "Beginning Remote Video Astronomy … So, what's the matter?".

I have struggled with what I should include in this book. Just dealing with the rapidly growing hobby of video assisted Astronomy could be a book unto itself, and there are excellent books that discuss this wide ranging topic. What I was primarily interested in, initially, was the ability to use video assisted Astronomy inside a warm car or my house, in a way that can be set up for an evening viewing session and then packed back up. So that is really what this book is about. I'll describe what I have found that works and doesn't work, and I am sure there are other good combinations of equipment than what I describe that also work well. New products are coming out all the time that will make Remote Video Astronomy even more practical and affordable.

This is a "Beginning" type book and does not get into how to set up a remote controlled observatory. Neither does it get into how to take high resolution images and spend hours of post processing time to produce amazing photos to show others. My goal was to explain how you could start using your telescope remotely without breaking the bank, and use the latest amazing video technology to see both solar system and deep sky objects while inside in near real time.

As I said in my other books, I hope you find some suggestions here that make your life easier in this great hobby of Astronomy!

Jim Meadows

August, 2014

Dedication

To my wife, Sharon, for putting up with my late nights, seeing all sorts of equipment going in and out of the house (also late at night), listening to me describe things that work and don't work, giving up a closet for my telescope stuff and patiently reading over my drafts.

Disclaimer

The information in this book is presented as accurately as possible based on the information the author had at the time of publication. The author makes no guarantees regarding the information contained in this book. However, the author has worked hard to ensure that the information is as accurate as possible at the time of publication. Equipment costs are approximate prices for Internet purchases. Please check system requirements for product compatibly with your equipment and/or computer before any purchase. The author does not receive income from any of the products described in this book.

Acknowledgements

Celestron and NexStar are registered trademarks of Celestron, LLC.

Orion is a registered trademark of Orion Telescope and Binoculars.

Apple is a trademark of Apple Inc.

All other proprietary names are the property of their respective companies.

Introduction

I am an engineer and like to solve problems. I also like Astronomy. As I have said in my other books, this is not a book about Astronomy though. There are many good books about beginning Astronomy and learning the night sky. This book addresses issues you may run into when starting to use your telescope and video assisted equipment remotely to see both solar system and deep sky objects while inside in near real time.

The examples in this book reference Celestron mounts I have used and specific video cameras such as the MallinCam cameras. Many items covered apply to other mounts and cameras as well.

This is not intended to be a replacement for your manuals. Be sure to follow the manual as much as possible. This book contains helpful information that is not detailed in the manuals and will hopefully save you some time and frustration.

The format of this book is simple. It contains a list of questions with a discussion of the issue at hand.

Beginning Remote Video Astronomy …

1. So, exactly what do you mean by "Beginning Remote Video Astronomy" (RVA)?

I thought it would be good to define what I plan to describe in this book, so I'll take each part of the title separately to answer this.

Beginning: The purpose of this book is to help you begin using your telescope in a new way. Instead of being at your telescope peering into an eyepiece, you will learn some basic methods that enable you to set up your telescope outside for an evening viewing of sky objects while inside. This is not a book about how to set up a permanent remote controlled observatory. Instead, you will learn how to configure your gear so that you go outside and set up, go back inside for your viewing session and later pack back up when done. You will need a beginning level of commitment to obtain reasonable equipment and master certain techniques to be successful in this endeavor.

Remote: For purposes of this book, remote means a distance of 15-100 feet from your telescope so you can be inside rather than having to be at your telescope. This could be inside your house, inside your vehicle or inside a tent or other protected viewing area. Typically a single cable can run from your telescope through a window or specially prepared opening to your inside area. Some Wi-Fi solutions are described where practical.

Video Astronomy: I use the term Video Astronomy to mean any method that provides near real time viewing through your telescope. Images on a screen are updated at regular intervals and may have some basic image enhancements applied while you watch. The imaging device can be a video camera producing signals that can be displayed directly on your TV, or other types of cameras that can provide image updates in a reasonable time frame such as some USB based cameras. The joy of Video Astronomy comes from seeing images in near real time without hours of post processing being required. Video Astronomy can make your telescope seem like a much larger telescope allowing you to see more color and detail than you ever saw through your eyepiece. You will be amazed at what is possible to be viewed from a light polluted backyard. Even dim remote objects are now assessable … in color! It is also much easier for this to be a shared experience since several people can look at an image on a screen at the same time!

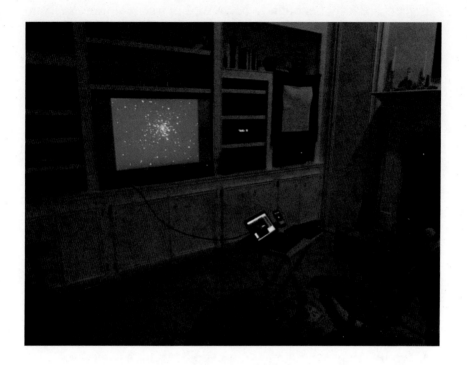

RVA: I use this as a shorthand abbreviation for Remote Video Astronomy, where Video Astronomy is combined with remote control and viewing techniques. Some cameras are suitable for Video Astronomy, but not for RVA. Some controls are good when near the telescope, but not for RVA. But when you have a good combination of Remote control and Video Astronomy, then RVA becomes a reality.

2. So, what is "near real time viewing"?

I consider "near real time viewing" to include cameras that can update your display anywhere from instantly to up to 3 minutes, depending upon the object you are viewing. Some objects are bright enough to be viewed in true real time, such as the moon and some planets. Other objects require exposure times of 1-10 seconds before your display is updated. Nebulae and other deep sky objects may range from 10-30 seconds, and some faint objects, such as the Horsehead Nebula, may require up to 3 minutes. Anything longer, and you begin feeling like you are just taking pictures and are no longer "viewing" objects. But unlike photo sessions of capturing images of just a few objects for later processing, Video Astronomy enables you to see several deep sky objects in a single evening, just like you would during a viewing session at your telescope.

3. So, what is the minimum I need to begin Remote Video Astronomy?

Assuming you already have a computerized GoTo type mount and telescope, the simplest way to begin is to insert a video imaging camera into your eyepiece with a cable long enough to reach inside, and extend your hand controller inside with an extension cable.

You could begin with a very low end video camera, but you will most likely be limited to only viewing the Moon and maybe a couple of planets. I recommend you at least begin with something that can carry you further so you can get the thrill of seeing deep sky objects as well. A basic focal reducer will cut down on frustration while trying to find objects by providing a wider field of view and will make your images brighter. A focus mask and small LCD monitor that you can use at your telescope will also help assure that you have a good focus for great images.

So, I suggest you start with the following equipment:

a. A good entry level video astronomy camera, such as the MallinCam Micro video camera ($100) and associated nosepiece, power adapter and 25 foot extension cable ($70).

b. A basic .5x focal reducer that screws onto your video camera, such as an Antares 1.25" 0.5x Focal Reducer ($35) or the MallinCam 0.5X focal reducer ($50).

c. A cable to extend your mount hand controller inside, such as the 25' AstroGadgets Celestron NexStar Hand Control Extension Cable ($15).

d. A focus mask that matches your telescope, such as a Farpoint Bahtinov Focus Mask ($15-$30)

e. A small LCD monitor, such as an Intsun 4.3" LCD Monitor ($20) and AC to DC 12V power supply adapter for Security Camera ($6) or 12V vehicle lighter adapter ($8).

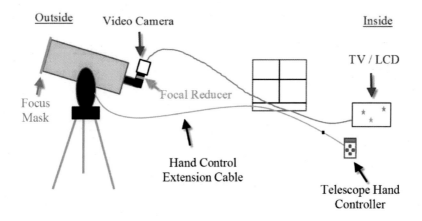

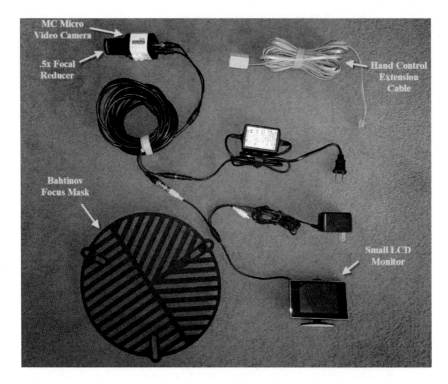

All of this together will be about $275 plus shipping to get you started.

4. **So, how do I use this basic setup to get "first light" inside?**

Perform the following to set up your equipment:
a. Unplug your hand controller from your mount, insert the hand controller extension cable into your mount and plug your hand controller into the end of the extension cable. You can use Velcro to bundle the excess cable until you are ready to extend it inside. You may want to also Velcro your hand controller to a suitable spot on your mount.

b. Set up your mount outside, turn it on and perform an alignment at the telescope in your normal manner using your hand controller, finder scope and eyepiece(s).

c. Select and slew to a bright star such as Vega or Capella, and use the hand controller to center it in the eyepiece so you can easily adjust the focus for your camera when you insert it.

d. Screw the nosepiece onto your camera and the .5x focal reducer onto the end of the nosepiece if you have not already done so.

e. Connect the cables to your video camera and the LCD monitor. Insert the video camera power cable and LCD power cable into power. Adjust your camera settings for initial coarse focusing. For the MallinCam Micro, press the buttons on the back of the camera while watching the LCD monitor to set the Lens exposure to x128, which is about 2 seconds. (See your manual for other recommended settings. The *Micro-Ex Camera* document listed in Appendix A is a great reference on how to setup and use the MalliCam Micro.)

f. Remove your eyepiece and insert the video camera.

g. If you have centered your target star well, you should see some light from the star on your display. The purpose of the 2 second camera setting is to be able to at least see the star since it can be fairly faint when out of focus. Turn your focus knob a little and pause for a couple of seconds to see if you are going in the right direction. If the image gets larger, then adjust the focus in the other direction. Continue to adjust the focus until the star is close to focus.

h. Now place the Bahtinov Focus Mask over the end of the telescope.

I prefer a focus mask that fits over the end of the telescope rather than inside the end of the telescope since it can also be used over the end of a dew shield as shown here. As you get close to focus, you can easily put this type of focus mask in place without disturbing anything and finish your focusing.

You will see lines extending out from the star. Continue to adjust the focus until the middle line is centered between the two diagonal lines.

You may be able to shorten your camera exposure (e.g. to X16) when doing this to see the results more quickly as you finish your focusing. Remove the focus mask when done. If you do not have a focus mask, you can achieve best focus by watching a nearby faint star and adjusting the focus until it is as bright as possible.

i. After you are satisfied with your camera's focus, adjust your camera exposure setting for what you plan to observe. For the moon or planets you will use exposures under a second. For nebulae and other deep sky objects I go ahead and set the camera to x128 or x256 (about 2 or 4 seconds).

j. You are now ready to move inside. Uncoil the hand controller extension cable and pass the hand controller and cable through an opening such as a raised window.

k. Do not unplug the hand controller since that would lose your alignment. You can disconnect the camera cable from the LCD and power in order to extend the cable inside. (At this point I put a towel along my window opening and lower the window back down). Before you go inside, check your cables at the telescope to make sure you have enough free play to accommodate your telescope slewing in any direction.

l. When inside, apply power again to your camera and connect it to the LCD or other viewing device such as your TV composite input.

m. You are now ready to start using your video astronomy setup remotely from inside. Use your extended hand controller inside as you normally would to select and slew to objects. Use the up/down/left/right buttons as needed to center objects on the screen.

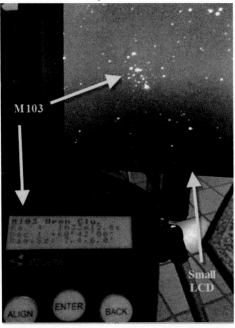

5. So, what are some tips as I get started?

Here are some helpful tips:

a. First perform a trial setup during the daytime to slew and focus on land objects without using a focus mask. Determine how much cable play you need for slewing and ways to secure the cables. Work with the menu system of your camera and learn how to change its settings as described in its manual. Also, it would probably be good to just stay at your telescope for your first few viewing sessions to learn how to use a video astronomy system before going remote.

b. If you can see your telescope through a window when you go inside, you may want to watch it as it performs a GoTo slew to ensure the cables do not get tangled or caught. If this is not possible, you can set up an inexpensive Wi-Fi camera to "watch" your telescope from inside. See question 7g.

c. Depending upon your camera settings, you may see star trails on your display as the telescope slews to its target. Once it is has finished slewing, you may see smaller star trails for a short period of time depending upon how much backlash you have in your gears. Once any gear slack is taken up as your mount begins to track, these star trails will disappear and become points of light.

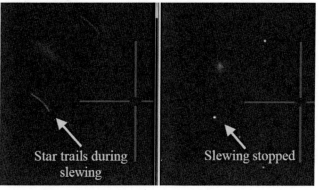

Star trails during slewing

Slewing stopped

d. If you need to use the hand controller to better center your object, use short presses to get a feel for how far one to two second button presses move objects in your field of view. You may see short star trails when you press a button depending upon your camera exposure setting. Be patient as you learn to work with the slight update delays and how to maximize your adjustments without overshooting.

e. Start with fairly bright, easily recognized star clusters such as the Hercules Cluster to learn how to center your object. Then move to a bright nebula such as the Orion Nebula.

f. If you use a battery power tank at your telescope, you can use adapters to power your video equipment from the same battery pack instead of using AC. See question 51.

g. Be sure to give yourself enough time to pack up when done. Leave the nosepiece and focal reducer on the camera to help protect its CCD sensor. Place a lens cap over its nosepiece and store the camera in a protective bag or case along with the LCD. Coil up your cables and store them so they don't get tangled. I secure mine with Velcro and put them into inexpensive plastic containers.

6. So, what can I do if objects do not appear in the field of view after a GoTo slew?

If the object does not appear in your field of view after a slew, your mount's GoTo pointing accuracy may not be well matched to your camera's field of view. This is particularly true of lower end Alt-Az mounts. Without a .5x focal reducer, your camera field of view is probably similar to an 8mm eyepiece. With a .5x focal reducer you will have a wider field of view. If you find you are still having trouble seeing your object after your mount slews to it, you may want to add another focal reducer to your telescope's visual back, such as the Celestron .63 focal reducer. See questions 31 and 32. This will greatly improve your mount's ability to place the object within the field of view and increase brightness, but it will also make objects smaller. This is beneficial though when viewing objects that extend across a wide field of view, such as some clusters and nebulae.

Viewing planets can be a particular challenge depending upon your mount's GoTo accuracy. You typically do not want to use focal reducers when viewing planets. It is even desirable to use a 2x-3x Barlow lens to increase the magnification for solar system objects (other than the Moon). Since you cannot be at your telescope to use your finder for assistance when centering an object, you could consider adding another camera as a remote finder scope to help center planets from inside. See question 12.

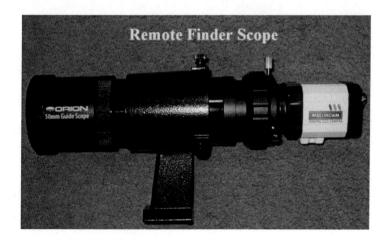

7. So, what else would be helpful for an expanded system?

Here are some helpful items you may want to consider when expanding your Remote Video Astronomy system:

a. SkyGlow filter. If your location is near a city, add a sky glow or light pollution filter onto the nosepiece of your camera (after the focal reducer), such as an Orion SkyGlow Eyepiece Filter ($63) or Celestron UHC/LPR Filter ($55).

It is actually amazing what you can see with a video astronomy system in spite of light pollution due to the camera's increased light sensitivity and ability to average out some of the background noise. A good sky glow filter will reduce the effects of these extraneous light culprits.

b. Wireless mount control and planetarium software. Instead of extending the hand controller inside, you can improve your remote control ability by getting a Wi-Fi device to control your mount such as Southern Stars SkyFi ($160), hand controller connecting cable ($30) and their hand held SkySafari Plus planetarium software ($15) for selecting and slewing to objects (SkySafari can be run on a smart phone, iPad or Mac). The Orion StarSeek Wi-Fi Telescope Control is essentially the same device and works with Orion's Star Seek Astronomy phone app. Celestron also has their own Wi-Fi device and planetarium app. The visual interface of these systems can greatly enhance your RVA experience.

c.   Remote camera control. If you use the MC Micro
     camera, you can add a Zengineering Micro-EX USB
     direct control cable ($57) enabling you to change the
     exposure times and other settings from your laptop inside
     without having to press buttons on the back of the
     camera. Remote camera control cables and software for
     higher end cameras like the MallinCam Extreme can
     provide a wide range of real time adjustments to your
     video image with easy settings for specific types of
     objects (Moon, planets, deeps sky objects, etc.).

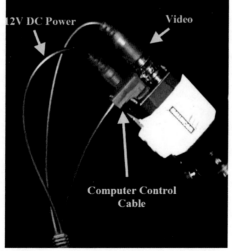

d. Small portable LCD-DVR. A small battery operated LCD-DVR such as the Orion StarShoot LCD-DVR ($200) is a handy multi-purpose device. You can use it at your telescope to assist with focusing your camera and then bring it inside to easily capture images without depending upon a computer. Due to its small screen, it is strictly a one- person-at-a-time viewing device, but the image is very crisp and clear. The Orion LCD-DVR is a well-made device that is popular for recording DSO images or planetary video clips to an SD card and includes a remote control for easily taking pictures/videos of what you are currently viewing.

e. USB Video capture device. You can plug your camera composite video output into a device such as the Orion Video Capture device ($43) that plugs into your laptop (PC or Mac) for near real time viewing on your screen and capturing images or video on your hard drive. They typically also have S-Video input for higher end video cameras that output S-Video.

The MallinCam MCV-1 Frame Grabber ($80) is a Video Capture device designed specifically for astronomy video capture.

f. F/6.3 rear cell focal reducer/corrector lens. This type of focal reducer screws onto the rear cell of an SCT telescope and reduces the focal ratio by a factor of 0.63 enabling a reduction in exposure time by 50% and provides a wider field of view. Both the Celestron F/6.3 Focal Reducer ($145) and the Meade F/6.3 Focal Reducer ($98) also provide corrections for improved edge of field images. I leave my F/6.3 focal reducer on my telescope most of the time except when viewing objects that require high magnification such as planets.

g. Wireless IP camera. This is not for viewing sky objects – it is for watching your telescope from inside as it slews to make sure all is OK outside. You can accomplish this with a fixed focus low light Wi-Fi camera such as the D-Link DCS-932L ($60) and an app running on your phone. Setting up the wireless camera with your home Wi-Fi allows you to view the telescope from your phone, tablet or computer while you stay inside slewing to various objects during your viewing session. For a little more

($75-$100) you can get a wireless IP camera that also supports pan and tilt.

h.   Battery packs. I have been able to make all my equipment portable and run off battery packs with appropriate cables.

You can use 12v lighter splitters to provide additional connections to a power tank for your camera, LCD etc. I have also found the small rechargeable battery packs such as the Anker models ($20-$50) useful for powering a wireless IP camera, your video camera, LCD, a hub and other devices.

i. <u>Expanded capability Video Astronomy camera(s)</u>.
After I became sold on how great Video Astronomy is, I
took the plunge and purchased the MallinCam Extreme
X2 ($1299-$1740). To sum up its abilities in one
word...Wow!

If you get to the point where you want an expanded
capability Video Astronomy camera but don't want to
spend this much, there is now available a MallinCam
Junior Pro PC camera ($600) that has many of the

features of the Extreme and includes a video/power cable plus a cable for computer control. When thinking of the cost of a video camera, just remember the camera is replacing all your eyepieces when you switch to Video Astronomy.

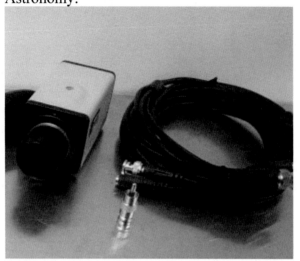

There are certainly other good Video Astronomy cameras on the market and I am sure we will see even more over time. I have mainly purchased MallinCam products because they have been the pioneer in this field for years.

8. So, can I adapt webcams for Video Astronomy?

If you have a webcam you no longer use, such as a Logitech 3000 or 4000, you can try adapting it for use with your telescope. You can get a nosepiece that can screw on in place of the web cam lens, such as the OPT 1.25" Web Camera Nosepiece ($15), enabling it to be inserted like an eyepiece into your telescope. I adapted an old Trust webcam using this technique. I first had to remove the side screws and the front cover to unscrew its lens so I could screw in the nosepiece as shown below.

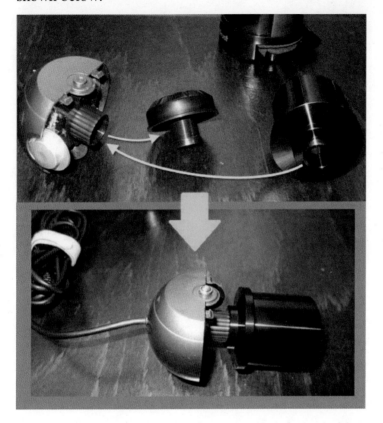

   If you do not have a web camera, it is probably not worth buying a new web cam and a nosepiece to adapt it. Instead you could just buy an entry level USB camera for Astronomy such as the Orion StarShoot USB Eyepiece Camera II ($56). You can use a USB extension cable to connect your webcam to your laptop inside for use with software that works with your camera.

With low end USB cameras, you will be limited to viewing mainly bright objects such as the moon and maybe some planets because you have limited control over the exposure and gain. Also, since you can only view the image on your laptop when using a USB camera, you will have to take your laptop outside to perform your initial focus before you run your cable inside.

For RVA, these types of low end USB webcams are really only suited for viewing the Moon or bright planets. So don't expect to be able to see nebulae and faint galaxies. Even trying to view Jupiter remotely could prove frustrating because you need to use a 2x or 3x Barlow to best view a planet at higher magnification. This makes it difficult to get a planet in the camera field of view unless you are at the telescope and can use your finder to first center your telescope on the planet, which defeats the purpose of being remote from the telescope. It is possible to capture short video clips of Jupiter or Saturn with webcam type cameras, and later post-process the file using software such as Registax to use only the best images in the video clip, align and stack the selected images and improve sharpness and contrast. But this is obviously not near real time viewing. So stick to just viewing the Moon remotely with these types of cameras to get a feel for what remote viewing is like.

But be warned, you will then want to see galaxies, nebulae and other deep sky objects from inside, and a better camera will be in your future soon.

9. So, can I adapt security cameras for Video Astronomy use?

Some security cameras are suitable for being adapted for Video Astronomy use. Modifying certain Samsung security cameras has been a popular way to have a camera with enough low light sensitivity and gain adjustment for video astronomy in the $200 range. I started my video astronomy with a Samsung that had its lens and filter removed and a nosepiece added. I viewed the video image produced by the Samsung using a small LCD at the telescope.

Like many video cameras, the Samsung camera settings are adjusted using the buttons on the back of the camera to navigate its menu system. It has the ability to set the exposure up to about 8 seconds (sense-up = 512x), enabling you to see nebulae and other deep sky objects.

I took the back off my Samsung and soldered cable connections to the push button switches and ran the cable out to a phone connector on the side.

I made a remote push button box using buttons from Radio shack with a 25 foot 6 conductor phone cable enabling me to control my camera settings from inside.

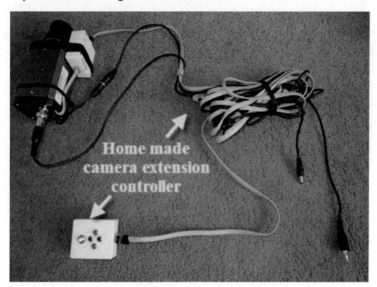

Home made
camera extension
controller

Unless you already own a security camera you can modify, now you are better off just purchasing the MallinCam Micro ($100) which has the capability of exposure settings up to 17 seconds. You will have much better results using the MC Micro since it is tailored specifically for Video Astronomy. You can also purchase a control cable for it in varying lengths from 25 - 100 feet that allows you to use its menu system remotely from your computer inside to change camera settings instead of having to press buttons on the back of the camera.

10. So, what are the various methods used to display images remotely in Video Astronomy?

When I think of cameras for Video Astronomy I put them into one of two broad groups – Analog and Digital. Now you may be thinking, why use analog cameras? Isn't digital better technology? Digital may be best when you are taking pictures, but remember, with Video Astronomy you are trying to achieve near real time viewing. In fact, some only consider analog video to be true Video Astronomy. As I have previously said though, I consider Video Astronomy to include anything that allows you to achieve near real time viewing. Let's look closer at these two types of technology.

Analog video cameras are classified as either NTSC or PAL based on the format and refresh rate of the analog video signal they produce. The term PAL is often used to refer to the 625-line/50 Hz (576i) television systems, to differentiate from the 525-line/60 Hz (480i) systems used with NTSC. Most analog video display devices used in the Americas and some parts of the world are NTSC while Europe uses PAL. Both of these typically have a composite video output using BNC or RCA type connectors, or S-Video if the camera supports it. Composite video is carried on a single channel while S-Video splits the signal into two channels reducing cross talk for a cleaner image. Another type called component video splits the signal into three channels requiring bulkier cables and typically is not used in Video Astronomy. The very low end analog video cameras only produce a basic live video image which is fine for bright objects such as the Moon or planets. Better analog video cameras used in Video Astronomy have a means of effectively accumulating the video image over a specified period of time, sometimes referred to as sense-up, to extend the time exposure capability enabling near real time viewing of Deep Sky Objects.

Digital Video has essentially replaced Analog Video for TV transmissions. Digital video cameras use cables such as USB or Ethernet cables to transfer the digital signal. Low end USB webcam type cameras typically provide live images only, with no means to provide multi-second exposures. Higher end digital cameras capture an image over a specified exposure period and transfer the final image over USB or Ethernet to a computer for displaying.

Both analog and digital cameras can be purchased as a black and white camera or a color camera. Color gives you a one shot color image capture. Black and white cameras can be great for producing high resolution pictures, but require the use of multiple exposures using different color filters to produce a final color image and thus are typically not used for RVA.

11. So, what are the pros and cons of the different types of cameras used in Video Astronomy?

Why use Analog cameras for Video Astronomy? Analog cameras provide a simple, effective means for near real time viewing. No computer is required. Just connect the analog video output to a display device and you are ready for viewing. The video signal output is instantly viewable. Even when a timed exposure is used, the resulting image is immediately seen when the exposure is finished.

Digital devices normally require a computer to receive and display the video image. For timed exposures, an additional 1-3 seconds may be required to transfer the image over the digital cable. If you were using 2 second exposures and it took 2 seconds to transfer the captured image you would see updates every 4 seconds. With Analog Video the updates would occur every 2 seconds, immediately after the exposure is complete. For longer exposure times, such as 1-3 minutes, the extra digital image transfer time is not as noticeable. Digital video can provide higher resolution images than analog, but the higher the resolution the longer it will take to transfer the image.

Analog video cameras require a separate power cable. For remote control of analog video cameras, a separate control cable is also required. For digital cameras, video camera exposure control is usually provided over the same USB or Ethernet cable used for the image transfer along with power, so only one cable is needed.

Here are some pros and cons of some specific types of cameras for use with Video Astronomy:

a. Analog Video eyepiece (Orion Star Shoot Video Eyepiece Camera). These cameras are easy to use and only require a video display, but are limited to very bright objects only, such as the Moon and some planets. They are inexpensive but have a lower quality image and have no timed exposure ability.

b. Standard analog video cameras with variable time exposure capability (Modified Samsung Security Cameras). These cameras only require a video display to use. Exposure time is usually set using push buttons on the back of the camera.

c. Astronomy analog video cameras with variable gain, time exposure and other settings (MallinCam Micro, Junior Pro, Extreme, etc.). Being able to tailor the gain, exposure and other settings to the type of object being viewed can provide significant improvements in your displayed image. Software is typically available to allow remote computer control of the camera settings and some real time processing to enhance the image.

d. Low end USB webcam type astronomy cameras (Orion Star Shoot Solar System camera). These provide some improvement over the basic analog eyepiece camera, but are still primarily limited to solar system objects.

e. USB based cameras with computer control of gain, exposure and other settings. Some are designed specifically for Video Astronomy (MallinCam Solar System Imager). Here is an image of Jupiter on my TV inside using the MC SSI and its control software.

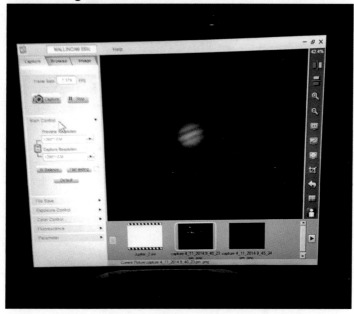

Others USB cameras are designed more for image capture and post-processing. (Orion Star Shoot G3 Deep Space Camera) The Orion G3 can be placed in a looping mode with color settings that provide near real time video updates for RVA use.

f. <u>High End Ethernet based cameras with computer control of gain, exposure and other settings.</u> (SBIG) These are often used for image capture and post processing.

g. <u>DSLR's with USB computer control and image transfer.</u> (Canon EOS). This is more an almost near real time viewing solution since you must manually initiate each image capture.

12. So, is there a way to use a finder scope remotely?

Sometimes you need to look through your finder to better center your target object for the camera attached to your telescope. But how do you do this when you are not at your telescope?

By adding to your telescope a small guide scope, such as the Orion mini 50mm Guide Scope ($80), and inserting a camera, you can have a remote finder scope to assist moving to and centering objects in your main camera's field of view. Orion also makes a mini 50mm Guide Scope with helical focus ($150) that makes it easy to adjust the focus for the camera. A MallinCam Micro can be used with the Orion Mini 50mm Guide Scope for an excellent remote finder.

You can even use crosshair overlays when using the MC Micro camera for a finder. See Appendix B for adjusting its settings as a finder camera. If both your main camera and finder camera are video cameras, see question 20 for more info on how to control and monitor two video cameras from one PC over one USB cable.

Another option is to use the Orion StarShoot AutoGuider SSAG ($280) with the Orion mini 50mm Guide Scope ($80) as a remote finder. You can get these two packaged together as the Orion Magnificent Mini AutoGuider ($350).

You will need to run a USB extension cable inside to your computer in order to "look" though your finder remotely. The SSAG has exposure times up to 10 seconds, allowing you to adjust the amount of star detail you see through the finder on your computer. The added advantage of the SSAG is that it can also be connected to your guide port and used as an autoguider.

If you add a remote finder, take the time during the day to focus and center an object using your main camera, and then focus and adjust the centering of your finder camera.

When you use it for the first time at night, perform finer focusing and centering of your finder camera using a bright star. See question 24 about how you can make your own 50mm focus mask to assist getting the best focus from your finder.

Having a remote finder means you can easily center and sync your software on a bright star in the region you are interested in from inside, and thus make your GoTo slews more accurate for that region to better ensure your target object will be in the field of view of your main camera.

13. So, what may be wrong when using a hand controller extension cable if my mount continues to slew after reaching its target?

Adding an extension cable to your hand controller can affect the voltage level or signal timing from the hand controller resulting in response errors when slewing if you do not have adequate power to the mount. If you find your mount not responding properly when you use a hand controller extension cable, check to make sure you have enough voltage to your mount by using a well charged battery or a good AC adapter.

I also found using my GPS plugged into the Aux port of my SE mount in the normal manner could result in a communications problem when using a hand controller extension cable. This sometimes causes the GoTo slew to not stop totally when it reaches the target object. When I disconnected the GPS module from the Aux port, my slews worked fine using the extended hand controller. Suspecting some timing issues between the Aux port and the HC port when using an HC extension cable, I inserted the GPS module in line with the extension cable by unplugging the GPS module from the Aux port and plugging it into the normal hand controller port on the arm. I then plugged the extension cable into the pass through port on the GPS and the HC onto the end of the extension cable as shown below. I strap the GPS onto the mount arm so all connections are near the normal hand controller port.

With this arrangement I could then use the GPS and the extended hand controller fine. It also makes it easier to remove the extension cable by simply unplugging it from the GPS module and then plugging the hand controller into the GPS module.

14. So, is it feasible to do Remote Video Astronomy wirelessly?

The answer is yes, depending upon how you define feasible. For example, it is possible to use low cost video transmitter/receivers to get your video inside without the use of cables, but you will probably be disappointed in the quality and reliability of the signal. You can use wireless USB hubs to extend control without using cables, but their bandwidth is limited so only certain equipment works well with them. You will have to spend some money for a wireless setup that works, but it is nice to just take your telescope outside, set it up and then go inside for viewing without having to run cables inside.

I had to limit myself to just the basics in order to keep wireless "feasible" for me (e.g. don't try autoguiding wirelessly). See questions 15-17 and 43 for more details on this and other wireless setups.

15. So, can I transmit the video to a receiver inside?

Transmitting video reliably is an interesting problem to solve. You can try to use composite video transmitter/receiver devices, but you will find outside to inside distances are limited, and signals are prone to interference since these devices must stay within prescribed power limits. I was able to use the IMAGE 2.4GHz Wireless Audio Video Transmitter and Receiver pair ($42) up to about 25 feet from the camera outside to the TV inside, which does fall into the lower end of my definition for "remote" operation.

To Camera

To TV

Outside

Inside

For longer remote distances, I found converting the camera video to a digital HDMI signal and transmitting the HDMI signal was a more reliable method. This technique takes advantage of emerging methods to stream HDMI signals wirelessly.

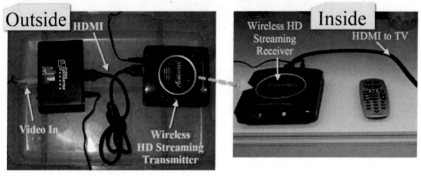

Outside HDMI

Inside

Wireless HD Streaming Receiver

HDMI to TV

Video In

Wireless HD Streaming Transmitter

I chose to use the Portta PETCSHP S-Video/Composite to HDMI Converter ($46) because it has inputs for either composite video (from my MallinCam Micro) or S-video (from my MallinCam Extreme). There are lower priced units you can get that convert only composite video to HDMI. Due to the growing popularity of using multiple HD TV's in the home, there are many solutions becoming available to allow you to wirelessly stream an HDMI signal to a TV in another room. I chose the Actiontec MyWirelessTV ($180) because of its distance specifications and its availability. If you get the Actiontec MyWirelessTV Wireless HDMI Kit from BestBuy ($200), it includes an IR repeater back to the transmitter that allows you to use a remote control from another room, which is convenient if you also want to use this equipment as a means to extend your cable/satellite TV signal to another room when not in use for RVA. This remote IR feature also allows me to use my Orion LCD-DVR remote control inside to capture images on the LCD-DVR located outside at the telescope.

16. So, how can I control my mount over Wi-Fi?

There are a couple of ways you can do this. If you already control your mount with planetarium software on your laptop through a USB connection to your Hand Controller, you could use an IOGEAR wireless USB Sharing Station to extend this wirelessly. You can also use a system designed for wireless control of the mount like SkyFi and SkySafari Pro as previously mentioned. See also question 41c.

17. So, how can I control my camera over Wi-Fi?

Camera control is optional, but nice to have. When you are at the telescope, you could preset your camera for the type of objects you plan on viewing. However, having the ability to adjust the exposure and other settings from inside is really helpful. As previously mentioned, I use the Micro remote camera control cable for my MallinCam Micro to adjust its settings from a laptop without having to press the buttons on the back of the camera. To convert this to wireless, I plug the remote camera control USB end into a IOGEAR Wireless 4-Port

USB Sharing Station GUWIP204. I use a laptop inside to connect to the wireless USB outside and then to the camera control's USB port.

18. So, what is involved in displaying the video on an HD TV?

Most HiDef TVs still have a Composite Video or S-Video input. If your TV has one of these, you can connect your camera output to the appropriate Video/S-Video input on your TV. If you TV does not have these optional inputs, or you would rather use HDMI, you can use a Video to HDMI converter like Portta PETCSHP S-Video/Composite to HDMI Converter as previously noted. The image will still be similar to using the Video/S-Video inputs, although some TV's perform some up-conversions that may appear to improve the image. Be aware that any video conversions will change the original image from the camera, so you are not exactly "seeing" the image as it is produced by the camera when converting it to HDMI.

19. So, can I adapt my DSLR for Video Astronomy use?

Yes, if your DSLR has the ability to connect to a PC for both control and transferring images. A DSLR can capture good viewing images in the 0-3 minute exposure that I feel is usable to qualify for Video Astronomy use. You will actually be taking a picture and viewing it on your computer, and can repeat the process fairly quickly, but it is not quite the same feel as cameras that continuously update their image at the end of each exposure time. A DSLR will typically have a much wider field of view than a camera made for Astronomy resulting in smaller images of your target object. You can increase the effective ISO to keep the exposure time down, but this will tend to show up any hot pixels your camera may have. Since you are taking pictures, you will be using disk space as you retrieve the image from the camera, but you can always later delete images you don't want to keep.

I purchased an almost new Canon EOS body from eBay to use for Astrophotography that I have also successfully used for Video Astronomy. It came with EOS software that allows me to control the camera exposure from a PC, take pictures and have them transferred to my PC. If you already have a DSLR with this capability, you can adapt it for use with your telescope by getting the appropriate T-Ring adapter for your camera and a 1/2 inch adapter that screws into it enabling you to insert your DSLR like an eyepiece. However I don't necessarily recommend you go buy a DSLR just for Video Astronomy since for the same money you could purchase a camera targeted for Video Astronomy.

You will probably want to use an eyepiece at the telescope to perform your initial alignment. Once your mount starts tracking, slew to a bright star and then remove your eyepiece and insert your DSLR and focus it. I find using a Bahtinov Focus mask on the telescope is very helpful for knowing when you have obtained good focus with your DSLR. If the object you use for focusing is very bright, you may be able to put your DSLR in live viewing mode to perform a rough focus. Then switch to exposure mode with a long enough exposure and ISO to capture a bright image of your target star so you can see the spikes produced by using the Focus mask. Alternate slightly changing the telescope's focus and taking a picture until the horizontal line appears evenly spaced between the other lines. I use a remote trigger at the telescope to take these images so I don't jostle the telescope when I initiate the exposure. I plug both the remote trigger and my camera USB cable in before I begin this process so I do not touch the camera again after it is in focus. I then run the USB cable inside and connect it to my laptop.

Once I slew to an object using one of the remote slewing methods such as SkyFi and SkySafari, I initiate an image capture of around 5 seconds to determine how well I can see my target object. If my target is not well centered, I make minor slew adjustments and take another image, and repeat until satisfied. I then increase or decrease the exposure time as needed to better see my target. When I select and slew to a new target, I just repeat this process. You will soon learn how to adapt your exposure setting to match your target type (nebulae, star clusters, etc).

Although I have not had good success trying to use USB video cameras over a wireless USB hub due to bandwidth limitations, I found I could use my Cannon DSLR with a wireless setup since it is not constantly transmitting images. By plugging the USB cable from the Cannon into an IOGEAR Wireless 4-Port USB Sharing Station, I can control the camera exposure and transfer images to my laptop over Wi-Fi. Combined with SkyFi+SkySafari telescope control, you can have a completely wireless setup.

20. So, how can I use video cameras for both my main camera and my remote finder from the same PC?

If you are at the telescope, a simple solution is to use a second monitor or a manual video switch to toggle between the video output of the two cameras. But with RVA, you are inside. You could run a separate video cable inside just for the finder view, but my preferred setup is a single USB cable from my outside equipment to my laptop inside. My first thought when I considered setting up my MallinCam Micro on my Orion 50mm finder scope was that I would add a second video capture device to my outside powered hub and use a second program like SmartCap for the view through the finder. When I attempted this, I ran into conflict problems with trying to use two video capture devices plugged into a hub connected to the PC, possibly due to driver conflicts and/or bandwidth limitations. I really wanted a single cable solution from one hub where I could manage and view both camera images from a single laptop. I found there was no problem using two USB-to-serial adapters on the same laptop to control the two cameras through their control cables, but having two video capture devices was a problem.

   I then realized I could use one video capture device for both cameras and just toggle between their images since my MallinCam Extreme X2 camera has an S-Video out connection and the MallinCam Micro output is Composite video.

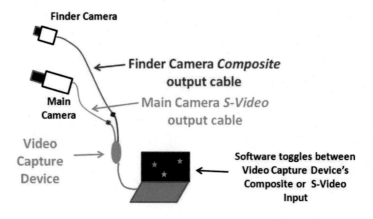

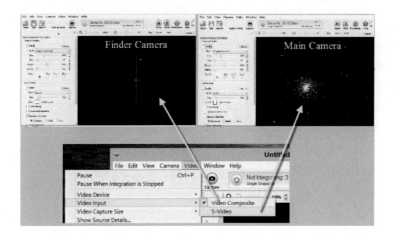

When using the MiloSlick software, I set the Video Input to Video Composite when I want to "look" through the finder camera to locate and center my target better before using my main camera. I then set the Video Input to S-Video to toggle to viewing the main camera image.

Even though I can use the same software to view either camera's image, the MiloSlick software is only controlling the MC Extreme X2 camera settings. I use the separate MallinCam Micro control application to adjust the MC Micro's camera exposure settings to better see the object through the finder as needed. You can use the MiloSlick software to toggle on/off a crosshair overlay on its displayed image as needed. When using the Micro as a finder, I prefer to configure its settings to display its own crosshair overlay as shown in the Finder View below. See question 21. Making use of this method keeps the dual camera solution simple, and you only need one video capture device and two applications (the MallinCam Extreme control/viewing application and the MallinCam Micro control application).

Using a finder camera can save time when trying to center DSO's since the finder can operate with a much shorter refresh rate. Below is where I used the finder to center a star cluster and then switched to the main camera view. I then changed the main camera to a longer exposure and the star cluster popped into view very nicely It would have been more time consuming if I were trying to center the star cluster using the main camera at the longer exposure.

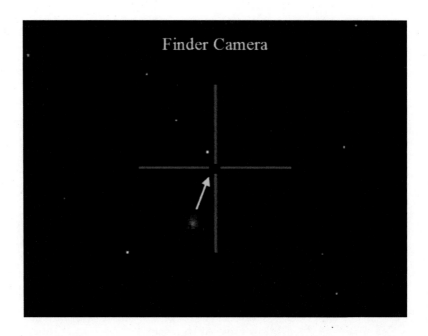

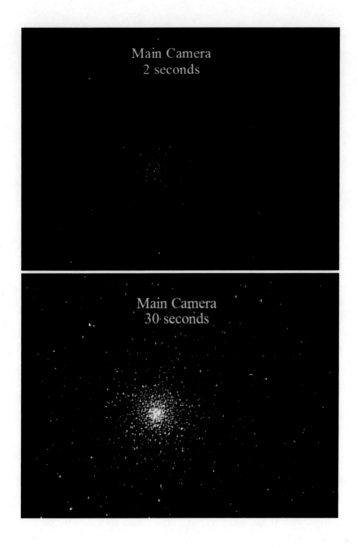

21. So, what do I need to consider when adjusting the settings on the MallinCam micro when used as a remote finder scope?

It is important to have a finder camera that supports adjusting both the gain and exposure as needed. I have found setting the

AGC to low and the Lens to X16 or X32 is a good setting for the MC Micro when used as a finder. You could get by with just setting the MC Micro to a fixed setting based on your anticipated targets for the evening, but it is much better to go ahead and get the MC Micro control cable ($57) and MC Micro control application (free) to allow you to adjust your finder camera remotely, based on the target you are acquiring.

The recommended settings in the manual for the MC Micro for the Moon and planets still apply when using it for a finder. For example, you should use Lens exposure settings instead of sense-up settings in order to make use of the special electronics that have been added for Video Astronomy. Using 3D noise suppression to stack 2-5 images helps smooth out background noise in the image, but does not help much with faint target detection. For Deep Sky Objects, turning AGC on by setting it to Low with Lens at X16 or higher works best for me. I sometimes make the Lens exposure time longer (X64 for faint objects) to "get my bearings" and see the target object, and then I can scale it back to X16 for centering, since screen updates are much faster at that setting. This is where having the ability to change the camera settings remotely from inside pays off. Be sure to also adjust the gamma setting to one that works best for the display you are using.

Normally, when using the MC Micro as the main camera for viewing through your telescope, the AGC should be set to "Off" to reduce amp glow and have a nice dark background. When using it as a finder camera, you are more concerned with locating objects rather than the viewing quality, so setting the AGC on to "Low" can really reduce the lens exposure time needed. This causes the background to no longer be dark, but the update rate is noticeably quicker which helps while you are centering the object in your finder. Once your target is centered in the finder, you can then toggle the video input to the main camera on your telescope with assurance the target will be in its field of view. Sometimes when I switch to my main camera view I do not even see my target object until I increase the exposure time. But when the first longer exposure finishes, it is great to see your target image pop onto the screen!

Crosshair overlays are helpful with a finder scope to provide consistency in centering your target. This can be done with a software crosshair overlay feature of the MiloSlick MallinCam Control application. Or you can make use of the MallinCam Micro Privacy settings to set up crosshairs in the camera image itself. See Appendix B for sample crosshairs settings for the MC Micro. It takes a little time to set up the crosshairs the first time, but then you can toggle them on and off as needed. When using the Micro as a finder, you could simply leave the crosshairs on all the time.

Below are images of M57 on a humid night near a city, first viewing through the finder using the MC Micro (with crosshairs off), and then switching to viewing through my C8 using the MC Extreme2.

22. So, what should I try out first at the telescope before operating remotely inside?

It would be good to get familiar with the basics of Video Astronomy before moving inside. Start outside during the daytime and focus your eyepiece on a distant object. Then remove the eyepiece and insert your camera and connect it so you can view its image outside either on a LCD screen, TV or your computer. Learn which direction you need to move your focuser to get a good focus. Try slewing the telescope to different positions and make sure your cables do not catch on anything. Learn how to change your camera settings and see their effects.

Now you are ready to try it out at night. Start again with the eyepiece to align your telescope and get your eyepiece in focus on a bright star. Then remove the eyepiece and insert your camera and connect it to your monitor or computer. Adjust your focus in the direction you learned during the day until you get the bright star in focus. If you have a Bahtinov Focus mask, place it on your telescope and adjust your fine focus. Remove the mask and slew to some other bright objects such as the moon or planets or bright stars. Learn how to use your telescope controls to center the image and adjust your camera settings as needed. Next slew to some of the brighter deep sky objects such as globular clusters (e.g. M13) and adjust your camera exposure to see more of the stars in the cluster. I enjoy looking at open clusters as well (e.g. NCG 7160) and learning how to adjust the camera to bring out the cluster formation and compare it to reference images of that cluster. Then try some of the brighter nebulae (e.g. M42) and learn how to adjust your camera settings to bring out the shape of the nebula and its colors. As you gain experience, try fainter nebulae (e.g. M27, M81, M82, etc.).

After you become comfortable with using your camera for Video Astronomy, add the equipment needed to control your mount and camera remotely (cables, powered hub, video capture device, USB adapters, laptop, controlling software, etc.) and set them up on a table outside beside your telescope. Practice slewing to objects and adjusting your camera settings using your computer. Watch your cables as your telescope slews and secure them as needed to keep the cables from catching on anything.

Once you have learned to control your mount with your computer while viewing outside, you are ready to move inside. Move your computer indoors and route your cables inside. Now get settled in to control your telescope and camera and view the heavens from inside. Start by slewing and centering bright objects again while inside. The Moon is great for target practice!

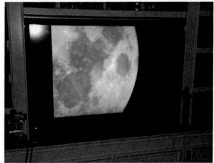

23. So, is there a way to make initial coarse focusing easier?

Sometimes it can be a challenge to even see a star using your video camera if your telescope is initially too far out of focus. Once you have it focused well, the next time you set up you can quickly fine tune your focus. But how do you get it focused that first time or when you add a focal reducer or Barlow lens? Here are some suggestions that make that first coarse focusing easier.

a. The first time you set up it helps to perform your first focus during the daytime when it is brighter and you can use a shorter exposure time. Pick an object in the distance and set your camera exposure as needed to begin to see an image. If it is all white, you probably need to cut down your exposure time (or put it on automatic). If you have first focused on a distant terrestrial object, you will be a lot closer to focus when you first try to focus at night on a star. During the day is the time to also make sure your finder scope is centered on the object your camera is seeing through the telescope.

b. If you have not performed an initial coarse focus during the day and it is now night, look for a bright local light such as a street lamp or other light in the distance. After you have focused on it, you can then slew to a star to perform finer focusing.

c. If it is night and there is no bright light nearby, slew to a bright star and center it in your finder. Adjust your camera exposure to x128 (2 seconds) or higher . If your telescope is not close to focus, this allows you to still see the star as a faint, large, out of focus object or ring. Move the focus some in one direction and wait for the image to update to see if the focus is better or worse. If the image is larger, then move the focus in the opposite direction. Continue to make incremental focus adjustments and check the image after it updates. As the star's image starts to become a bright point, you can reduce your exposure time to see your focusing results more quickly.

d. If you still have trouble even seeing an image using a longer exposure, you may need to swap out your camera with an eyepiece to make sure you have the star well centered in your field of view. Then re-insert the camera and adjust the exposure until you can see some image from the star to begin your coarse adjustment.

e. Once you have the star focused as well as you can, look for a faint star in your field of view and fine tune the focus further until the faint star is at its brightest.

f. If you have a Bahtinov Focus mask, place it on your telescope as soon as you get a fairly bright image of the star. See question 4h.

Then continue to adjust the focus until the center line is evenly spaced between the other two lines. Using a focus mask takes the guess work out of knowing when you have the best focus possible.

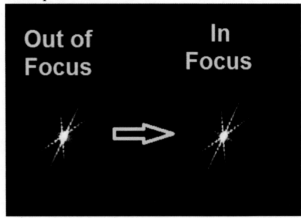

24. So, is there a way to make focusing a video finder scope easier?

Placing a Bahtinov Focus mask over the end of the finder scope would make it much easier to determine when you have the best focus for your video finder scope. However, I couldn't find a focus mask for a 50mm guide scope. So I made my own.

I first found a picture of a Bahtinov Focus mask on the Internet that when printed was just under 2 inches. I cut it out and taped it onto black construction paper. I then used an X-Acto knife to carefully cut out the gaps between the lines of the picture of the focus mask. I then cut around the outside of the image and separated the picture from the black construction paper mask cut out. I used a 1 inch strip of black construction paper and formed a short cylinder that matched the outer edge of the mask I had made, and used black electrician tape to attach them together. I also covered the outside of the small cylinder with black tape as well. The result was a small Bahtinov Focus mask that slips over the end of my 50mm guide scope!

My Focus Mask on Finder

When I need to make sure my video finder scope is in good focus, I slew to a bright star and adjust it to be slightly off center so it is not hidden by the crosshair. I then slip my focus mask on the finder scope and watch the display as I adjust the finder's focus until the middle line is centered between the two diagonal lines.

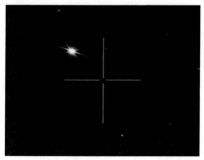

I later found you could print your own custom Bahtinov Mask at astrojargon.net/MaskGen.aspx by entering your finder scope's focal length and aperture and adjusting the Bahtinov factor (e.g. focal length 164, aperture 50, Bahtinov factor 50).

25. So, would a flip mirror be useful for RVA?

Although a flip mirror can be helpful when imaging at the telescope, its usefulness for Remote Video Astronomy is limited. When at the telescope, a flip mirror can be used to "share the view" through the telescope between an eyepiece and a camera. After you slew to an object, you can look through the eyepiece to initially see your object and center it, then you can flip the mirror up to allow the light to pass directly through to the camera for imaging.

Since you are not at the telescope there is no benefit to this arrangement for RVA use. Even if you were to use a second camera rather than an eyepiece, you would have to use an expensive flip mirror that can be controlled remotely. A reason you might want to use two cameras is so one could have a wide field of view for centering your object and the other for a better view of your target object. This would mean using different focal reducers on the two cameras, which in turn makes coordinating their positions so they are both in focus somewhat problematic. I found using a remote finder scope a better solution. See question 12.

26. So, if I can't turn the focus knob enough, what do I do?

When you use focal reducers or a Barlow lens with a camera, some configurations do not allow you to focus inward enough or outward enough to achieve good focus. For example, if you have a high amount of focal reduction and your camera is inserted into your diagonal, you may not be able to focus inward enough. Try removing your diagonal and inserting your camera in its place, which positions your camera closer to your telescope. You can also try using less focal reduction on your camera. If you determine you cannot focus outward enough, you can try adding extensions between your camera and the telescope. Generally speaking, the closer you can have your camera to your telescope, the better.

27. So, what do you typically use for your RVA system?

As I expanded my RVA system, I have found having a powered hub at my telescope as an interface to all my equipment is a great way to need only a single cable run into my house or vehicle for remote viewing. Here is what I typically use:

a. A Celestron CG-5 Advanced GoTo Equatorial Mount with a polar scope installed.
b. A Celestron 8" SCT.
c. A powered hub at the telescope.
d. A MallinCam Extreme X2 Deluxe (class 0 CCD) Video Camera inserted directly into the Celestron telescope (no diagonal) with:
    i. A MallinCam FR5 focal reducer nosepiece attached.
    ii. The X2 S-Video output connected to a video capture device which is plugged into the powered hub at the telescope.
    iii. The X2 MallinCam Control cable connected to a serial to USB adapter which is plugged into the powered hub at the telescope.
    iv. The X2 Composite Video output connected to the Orion LCD-DVR mounted on the telescope for focusing / viewing at the telescope.
e. A MallinCam Micro Video Camera inserted directly into an Orion Mini 50mm Guide Scope mounted on the telescope as my remote finder with:
    i. The composite output connected to the same video capture device used with the X2. (I toggle between S-Video and Composite Video to see the view through the telescope or finder scope.).
    ii. The MallinCam Micro Control cable connected to a serial to USB adapter

which is plugged into the powered hub at the telescope.

f. A Celestron StarSense Camera and Hand Controller (with splitter for the CG-5) with a serial cable connected to the Hand Controller and a serial USB adapter connected to the powered hub.

g. A dew shield for the Celestron 8" SCT with built in dew heater connected to a dew heater control module at the telescope.

h. A Microsoft Surface Pro inside my home connected to the powered hub outside using an active USB extension cable and running:

    i. MiloSlick MallinCam Control Software.

    ii. MallinCam Micro Control Software.

    iii. Starry Night with Celestron ASCOM support.

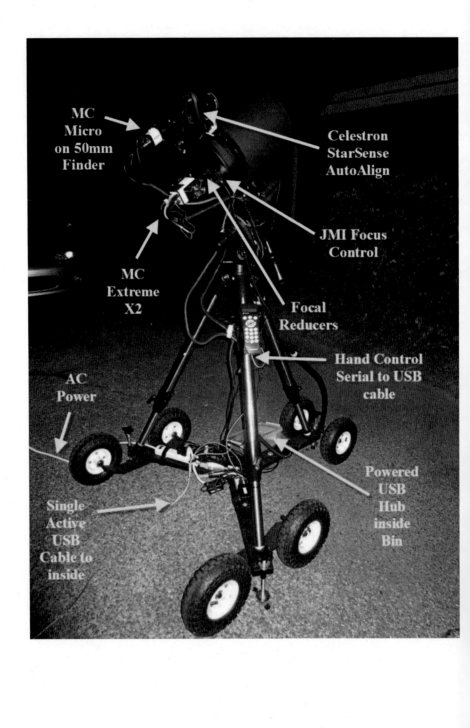

I keep my telescope and mount on a set of wheels with equipment attached and covered in my garage so all I have to do is roll it out to a marked location on my driveway.

I remove the telescope cover and perform a rough polar align with the polar scope. I power everything on and initiate a StarSense auto align. While it performs an automatic alignment, I connect an active USB extension cable to my powered hub and run it inside to my laptop. I then go back outside. When the auto align is complete, I slew to a bright star and adjust the focus of the main camera if needed. I am then ready to go back inside for my remote viewing session using Starry Night on my laptop to select and slew my telescope to my target objects.

Alternate setups I sometimes use include:

a. Adding a Celestron .63 focal reducer to the back of the telescope.
b. Using a JMI motorized focus control for my Celestron telescope connected to a PCFC USB control module plugged into the powered hub at the telescope.
c. Replacing the remote finder scope with an Orion Magnificent Mini AutoGuider mounted on the telescope connected to the CG-5 guide port and USB connected to the powered hub at the telescope.
d. Connecting the Hand Controller to SkyFi instead of a USB adapter and using SkySafari on my iPhone/iPad inside to slew my telescope to my target objects.

I can use the Orion Mini AutoGuider with PHD software for my finder scope and sometimes engage the autoguiding if I use longer exposures.

28. So, what <u>does not</u> work very well for Remote Video Astronomy?

Below is a list of things I have found did not work very well for RVA (they may work fine for other uses, just not for Remote Video Astronomy). Some of these I have already mentioned in other questions.

a. <u>Web cams</u> – typically they do not have any long exposure capability and are not sensitive enough. They can be fun to play with initially and OK for viewing the moon.

b. <u>Low cost eyepiece video cameras</u> – they can only be used for very bright objects such as the moon and some planets since they do not have any adjustable settings.

c. <u>Phone cameras</u> – battery life is not suitable for sustained video and remote viewing is difficult. (I did try AirPlay mirroring using Apple TV inside my house). You can use an adapter to keep it stable above an eyepiece for photos.

d. <u>Cameras that do not allow Automatic Gain Control adjustments</u> – the target objects will become saturated due to the dark background if you cannot limit the gain.

e. <u>Cameras that cannot be controlled remotely</u> – some that are good for Video Astronomy cannot be used for RVA since you cannot adjust settings remotely.

f. <u>Trying to use two USB video capture devices at the same time over a single USB hub connection</u> – you can have video conflicts and bandwidth issues.

g. <u>Trying to use any USB video capture device over a wireless USB hub</u> – there is not enough bandwidth.

h. <u>Manual flip mirrors</u> – not useful for remote operation for beginning RVA.

i. <u>Trying to use a camera without any focal reducers with a mount having limited GoTo</u>

accuracy – objects tend to not be in the field of view and will be hard to find (unless you also use a remote finder).

j. Trying to view planets with a mount having limited GoTo accuracy – it will be hard to get a planet in your viewfinder remotely (unless you also use a remote finder). A Barlow lens is normally needed for good views of planets which make GoTo accuracy even more important.

k. Using too much focal reduction – edges of images will be dark or distorted and it can cause reflections of bright stars.

l. Trying to extend the mount hand controller over 25 feet - voltage level and signal timing degrade over distance.

m. Trying to extend USB cables by using more regular USB cables – use only one or two long active USB extension cables instead.

29. So, what are hot pixels, warm pixels and amp glow and what can I do about them?

All imaging devices are prone to have some pixels that just do not cooperate with your desire to have a perfect image. Hot pixels are those pixels that are stuck on all the time and almost look like pin point stars, until you notice that all your images have these stars in the same spot on the screen. Higher quality cameras have very few hot pixels. A Class 0 sensor will have only a couple or none at all. Warm pixels are those that show up as the sensor becomes warm. There are techniques, such as using a Peltier thermoelectric cooler (TEC), that can reduce the effects of warm pixels. Real time processing performed by applications such as MiloSlick MallinCam Control software can be used to perform real time elimination of hot pixels before images are displayed.

Amp glow occurs as the electronics near the camera sensor heat up and affect a portion of the camera image. For example, it could appear as a lighter area in the upper left of the image. Thermoelectric cooling is useful to reduce this effect. The MiloSlick MallinCam Control software can also be used to reduce/eliminate amp glow as images are displayed. It also has the ability to apply dark frame subtraction in real time in proportion to exposure time if you have previously taken a dark frame.

30. So, how do you compare using a camera to an eyepiece's magnification and field of view?

This is an interesting question that can lead to debates and confusion, especially when discussing magnification power. The short answer to "how can you compare the camera's magnification to an eyepiece's magnification" is ... you can't! You can however, compare your camera's field of view to your favorite eyepiece's field of view.

Why can't you determine your camera's magnification? When someone asks "what power is your telescope", they are thinking about how big it makes something look as you peer into your eyepiece. This is dependent on two things – your telescope and your eyepiece. Everyone positions their eye to the same place when looking through an eyepiece and the size of the object projected onto your eye can be computed. The magnification can be computed by dividing the telescope focal length by the eyepiece focal length, in millimeters (Magnification = Telescope focal length ÷ Eyepiece focal length). For example, if you use a telescope of 1000mm focal length with a 20mm eyepiece, the magnification would be 50x (1000/20 = 50x). You could double that power by using a 10mm eyepiece (1000/10 = 100x). There are practical limits though on how much you can increase the magnification power of a given telescope.

With a camera, the image is displayed on a small LCD, or a computer screen, or a big TV screen, or whatever. The bigger the display, the bigger the object looks. Or the closer you move to the display screen, the bigger the object looks to you. So this is why it really doesn't make sense to talk about a camera's magnification power. It may not look as sharp when shown on a very big screen, but now you are talking about its resolution, not magnification. Resolution depends upon the pixel size of the sensors.

On the other hand, the area of sky observed when looking at your camera's image does not change when viewed on a small LCD or a large screen TV. So you can compare the field of view you see through a specific eyepiece to what you see on a display screen. The field of view for an eyepiece is the circle of sky visible through the eyepiece. The field of view for a camera is the rectangular area of sky seen when viewing the camera's image on a screen. The field of view through an eyepiece depends upon the telescope focal length, the eyepiece focal length and the eyepiece apparent field of view (AFOV). The true field of view seen through an eyepiece is its AFOV divided by the Magnification (FOV = AFOV ÷ Magnification). If we have an 8mm eyepiece with 50 degrees AFOV and a 1000mm telescope, then:

Magnification = 1000/8 = 125x
FOV = 50/125 = .4 degrees

But how do you compute the field of view for your camera to compare it to this eyepiece? You start by computing the Scale of each pixel, which is its sensor's pixel size in microns divided by the focal length of the telescope in mm times 206 (Scale = pixel size ÷ telescope focal length x 206). The field of view is then computed by multiplying the Scale of each pixel by the number of pixels (FOV ArcSec = Scale x #pixels). So if you have a camera with 8.4 micron wide pixels and 768 usable pixels across and a 1000mm telescope, then:

Scale = 8.4/1000 x 206 = 1.73
FOV = 1.73 x 768 = 1329 ArcSec
    = .37 Degrees (1329/60/60)

So, if you were to remove your 8mm eyepiece and replace it with this camera, you would see about the same area of the sky across as you see when looking through the eyepiece. Note that you should use the number of active pixels of the sensor when computing the FOV, not the total number of pixels since not all the pixels are used.

31. So, what are focal reducers and why would I use them?

Many deep sky objects are better seen with a wider field of view. You can expand the field of view of your camera by adding a focal reducer, which effectively reduces the focal length of your telescope. This also increases its effective light gathering power. If you were to screw a 0.5 focal reducer onto the nosepiece of the camera described in question 30, it would increase its field of view across to .74 degrees (.37 / .5), which would be similar to using a 50 degree AFOV 16mm eyepiece on that telescope. If it was an f/10 telescope, it effectively converts it to a faster f/5 telescope. This reduces the exposure time required to view deep sky objects and thus results in faster screen updates (i.e. even closer to real time viewing).

Instead of placing a focal reducer on the camera, you can attach one to the back of your telescope. I use the Celestron .63 focal reducer on the back of my C8, shortening its effective focal length by 37%, converting the f/10 optical system to a faster f/6.3 for a wider field of view and shortening exposures by 2.5 times.

You can also use multiple focal reducers to achieve even greater FOVs and faster update times, but only up to a limit. I have used a .63 focal reducer on the back of my Celestron C8 telescope and a .5 focal reducer on the nosepiece of my camera to achieve a total .315 focal reduction.

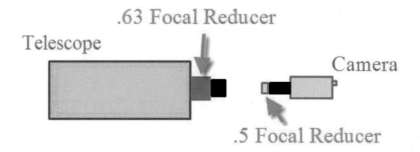

MallinCam makes a dual lens focal reducer (MC FR5) that can be used in combination with spacers on the camera to achieve a variety of focal reductions.

32. So, where should I put focal reducers in the optical train to avoid problems with focus, vignetting, etc.?

If you use too much focal reduction, you may experience vignetting, a circular fading of the image toward the edge of the field of view, producing a tunnel effect.

It can also result in a loss of focus around the edge of your image. You may, in addition, begin to notice reflections from bright stars in your image. If this occurs, you may need to adjust how and where you place your focal reducers in your optical train.

I like using the Celestron .63 focal reducer with my C8 telescope because it places the focal reduction nearer to the telescope, and it also includes field correction optics for better edge to edge focus. You remove your visual back from your telescope and attach the focal reducer in its place.

You can then attach your visual back onto the focal reducer. I prefer to use a Baader Click-Lock Eyepiece Adapter/Visual Back instead which allows me to securely attach my camera without a diagonal.

I have found it is better to put the camera as close as possible to the telescope when using focal reducers, so I avoid using a diagonal if possible.

If you use a multiple lens focal reducer attached to the camera, you may need to use smaller spacers to reduce reflections and vignetting if they occur.

33. So, when should I use a Barlow lens instead of a focal reducer?

While focal reducers are used to increase your field of view to see a wider area of the sky, Barlow lens can be used when you are viewing a small specific object such as a planet. A 2x or 3x Barlow lens will double or triple the image size of a planet on your display. Even though a Barlow lens will reduce the amount of light to the camera, your target is usually a bright object such as a planet where very short exposure times have to be used anyway to keep the image from saturating.

Since your field of view is expanded when using focal reducers, they make it easier to find and center your target object. With a Barlow lens, the field of view is smaller and you may not initially see your target in your field of view at all. A remote finder is a good companion when using a Barlow lens with your main camera. This will allow you to center your target using the remote finder so that when you switch to your main camera your target will be in its field of view. You will need to use very short slew adjustments when centering an image with a Barlow lens.

34. So, does Remote Video Astronomy require an equatorial mount and/or autoguiding?

The short answer is no, neither are required for Remote Video Astronomy. You can locate and view deep sky objects with an Alt-Az computerized mount if you perform a good alignment. In fact, using a Video camera while performing an alignment can result in a better alignment since you can more precisely center your alignment stars while viewing an image on a screen. If your GoTo slews do not bring your target into the field of view, you can add a remote finder to better center the target after the slew, and then you should be able find the object in your main camera field of view.

That being said, I have found a well aligned equatorial mount can slew my target into the main camera field of view without the need for a finder. (A remote finder will still help with quicker follow-up fine centering of your target in your field of view though).

Some Alt-Az mounts may not not track as well as an equatorial mount. Alt-Az mounts will also begin to be subject to field rotation blurring when using exposure times over 30 seconds due to the way Alt-Az mounts move as they track the night sky. However, there are many deep sky objects that can be seen using RVA on Alt-Az mounts with reasonably short exposures. An equatorial mount will allow you to better view fainter objects that require 30 second or more exposure times.

Due to the shorter exposure time used for Video Astronomy, autoguiding is normally not required unless you are viewing really faint objects or use special filters that require you to use exposures of several minutes.

35. So, what is the difference between GoTo accuracy versus tracking accuracy?

GoTo accuracy depends upon a good initial alignment. If you have a good initial alignment, your GoTo's can slew your target object into your field of view. Tracking accuracy, on the other hand, depends upon how well your mount properly tracks the sky after a GoTo slew. A lower end Alt-Az mount may be able to slew well to objects, but it may not continue to track them causing them to drift out of your field of view over time.

If an equatorial mount is polar aligned well and you perform a good initial alignment, it should slew your target into your field of view and keep it there over time as the telescope rotates about its polar axis to counter the Earth's rotation. An equatorial mount can perform good GoTo slews even if it is not polar aligned very well. If it is not polar aligned well, it just means it won't track the object very well after it slews to it. You will initially be able to see the object, but it will drift in your field of view. I use a polar scope to manually polar align my equatorial mount before I turn it on, and then perform a good initial alignment. I don't do a further detailed polar align unless I plan on needing additional tracking accuracy for longer exposures.

36. So, what is the difference between camera integration and frame stacking in software?

Cameras for Video Astronomy provide a means to electronically perform exposures longer than the normal refresh rate of the camera. This allows you to observe faint deep sky objects through variable exposure control, usually referred to as sense-up or in some cases a lens setting. These settings are usually in factors of two (X2, X4, X8, X16, etc) each doubling the time. If it is a NTSC camera at 60 fps, an exposure of X64 will be just over one second and X128 will be just over two seconds (128/60 = 2.133). Some cameras have additional modes to perform multiple second/minute effective exposure times. The MallinCam Micro can also stack up to 5 frames in the camera in a special 3D noise reduction mode to smooth out the background image.

Frame stacking in software takes each frame produced by the camera and adds the frames together. This can either be done in real time for Video Astronomy or later as a post-processing method to produce a single image of an object. Some frame stacking software can also align the images from the camera together using reference points before stacking the images so that a composite image can be produced even if the target drifts across the screen over time. This is most often performed in a non-Video Astronomy environment on a captured video clip that is post-processed to align and stack the images. It is particularly useful when high magnification was used, reducing the atmospheric effects of image jitter and minor changes in/out of focus. The software can eliminate out of focus frames and only stack the best frames to produce a single final image.

The MiloSlick MallinCam Control software can be used in Video Astronomy to perform limited stacking accumulation or averaging in real time of a specified number of frames to enhance the near real time image you are viewing.

37. So, can I view a variety of selections from the moon, sun and planets to deep sky objects such as nebulae and galaxies with Video Astronomy?

Yes. With the proper equipment, Video Astronomy is not limited to just bright objects. You will set up your equipment differently, however, depending on whether you are going to view the planets, moon or deep sky objects. When viewing planets you will want to use a Barlow lens to increase the size of their image for better viewing. When viewing the Moon, you typically will use the camera without any additional lenses. When viewing deep sky objects, adding focal reducers allows you to see a wider field of view and increase the effective light gathering power for viewing faint objects. These types of changes will require refocusing at a minimum and possibly performing an alignment after changing out the lens. So it is best to group your targets into these categories for a RVA session.

For the moon and planets you will set the camera to use shutter speeds under one second (e.g. 1/1000). For Deep Sky Objects you will set the sense-up (or Lens) to a setting like x128. If you are using a MallinCam with hyper mode capability, you can further adjust your exposure time from 3 seconds to multiple minutes for faint DSO's.

If you are interested in the sun, you can use RVA for solar observing during the day while staying inside where it is cool. YOU SHOULD ALWAYS USE A PROPER SOLAR FILTER OVER THE END OF YOUR TELESOPE WHEN IT IS POINTED AT THE SUN TO AVOID PERMANENT DAMAGE TO YOUR EYES OR CAMERAS. I also recommend removing your finder scope for solar observing sessions (unless you also place a solar filter on your finder scope). An inexpensive handy accessory that makes it easy to aim the telescope at the sun is the Tele Vue Sol-Searcher Solar Finder, which you can Velcro to the top of your telescope tube and watch (looking away from the sun) as you point your telescope.

38. So, when would I need different thread adapters (T-thread, C-thread, T-Ring, etc.) and nosepieces (1.25" and 2")?

Many video cameras have a C-thread mount for attaching lenses. To use them with your telescope you will need a C-Mount to 1.25" eyepiece adapter so you can insert them in place of an eyepiece (Or a C-Mount to 2" eyepiece adapter if you telescope uses 2" eyepieces). For more secure mounting of your camera to your telescope, you can use a locking eyepiece adapter like the Baader Planetarium Click-Lock Eyepiece Clamp/Visual Back which attaches directly to the visual back of your telescope.

If you want to use a DSLR in place of an eyepiece, you will need a universal 1.25" T-adapter (or 2") and a T-ring for your specific camera (Canon, Nikon, etc). You attach the 1.25" T-adapter to your camera using the T-Ring for that camera. You can then insert your camera into your diagonal or in place of your diagonal. If you do not need to easily swap your camera out with other cameras or eyepieces, you can just use the T-Ring for your camera to attach it directly to a T-mount for your telescope. If you have a Celestron Schmidt-Cassegrain, you would use the T-Adapter-SC which threads directly on to the rear cell of your telescope and thread the Camera with T-Ring onto it.

39. So, what basic filters should I use for video Astronomy?

If you live near a town or city, you can thread a sky glow or light pollution filter onto the end of the nosepiece of your camera to reduce the effects of reflected lighting, such as the Celestron 94123 1.25-Inch UHC/LPR Filter. Having a filter in place on the nosepiece at all times has the added benefit of protecting the inside of your camera and sensor from dust and other particles. If you attach a filter to your camera's nosepiece while inside in a low humidity environment, it will reduce the amount of moisture in the camera around the sensor.

Video Astronomy combines the best aspects of live observing and imaging, including the use of filters used for astrophotography. CCD sensors are typically sensitive into the near-infrared regions. An IR-cut filter can reduce the unwanted effects of this and help provide a sharper image, especially if you are using a refractor. An Oxygen III filter reduces glare and increases contrast and is good for viewing nebulae such as the Orion Nebula and Ring Nebula. A Hydrogen Beta filter is often used for viewing very faint trophy Nebulae such as the Horsehead Nebula. To view the Horsehead Nebula, you will probably also need a focal reducer with a 2-3 minute exposure using this filter.

You will need to plan your viewing session when using special filters with RVA since you will not be at your telescope.

Using a filter wheel makes it easier to switch between a few selected filters or no filter, but it will require you to make a trip outside to rotate a different filter in place (unless you spend big bucks for an automated filter wheel). See the Appendix for references on using filters for Video Astronomy.

40. So, what are the different ways a camera can be controlled from inside?

For RVA, it is best to be able to control your camera settings remotely. The method used to do this will vary from camera to camera. Some cameras provide a simple means to remotely use the On Screen Display (OSD) menu. Others use computer applications to directly change settings through a graphical user interface. Here are some of the methods you may run across:

    a. <u>A cable from the camera to a hand controller with buttons similar to those on the back of the camera allowing you to navigate the OSD menu and change settings without having to touch the camera</u>. Some cameras come with one or they can be purchased as an accessory.

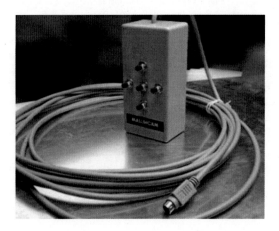

b. A wireless hand controller that controls specific camera settings through a small transceiver connected to the camera.

c. A cable from the camera to a USB connection to your computer running an application with buttons similar to those on the back of the camera to navigate the OSD menu and change settings while inside using your computer. This is how I computer control the MC Micro remotely.

d. <u>A cable from the camera to a USB connection to your computer running an application with an graphical interface to make it easy to set the camera to various modes and adjust settings.</u> This is how I control my Canon DSLR or my MC Extreme X2.

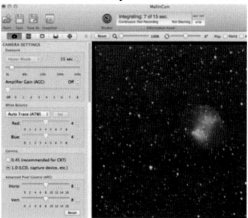

I have found I can use a wireless USB hub, such as the IOGEAR Wireless 4-Port USB Sharing Station, with cameras that have USB control cables, as long as the USB is not also used for the video image. By installing drivers on your laptop, you can connect to these USB devices over Wi-Fi and run applications as if the cameras was actually directly connected to your computer.

41. So, what are the different ways I can control my mount from inside?

For RVA you must be able to control your mount remotely. The method used to do this will vary from mount to mount. Hand controllers for some mounts can be used with extension cables for remote use. Some mounts can be connected to your computer by USB to allow your mount to be controlled by applications running on your computer. Some also allow you to control your mount from a computer, phone or tablet over Wi-Fi. Here are some of the methods you may run across:

a. A simple cable extension allowing you to use your Hand Controller remotely. I have used AstroGadgets NexStar Hand Control 25' Extension Cable to slew my telescope to selected targets from inside. This requires you to use the Hand Controller at the telescope for the alignment and then move it inside (without disconnecting it) to remotely control the mount. However, if you also have other devices such as GPS connected to your mount, then you may run into communication timing issues with this method. See question 13.

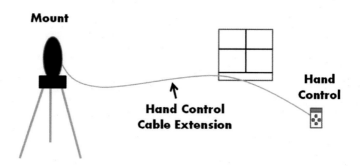

b. Attaching a serial cable to the hand controller and connecting it to a computer using a serial to USB adapter with a USB extension cable and running applications on your computer that can control your mount. Typically you would align your telescope using your hand controller at the telescope and then move inside and use an application on your computer to slew your telescope to selected objects such as Starry Night or other planetarium software that is ASCOM compatible.

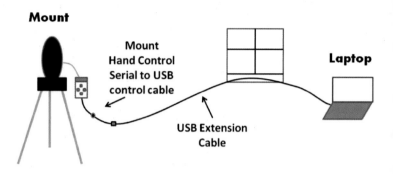

I have also used the IOGEAR Wireless 4-Port USB Sharing Station to do this over Wi-Fi.

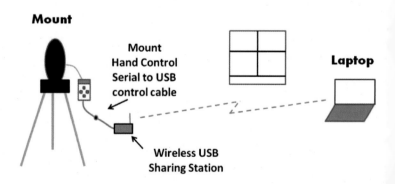

c. Attaching a serial cable to the hand controller and connecting it to a dedicated Wi-Fi device that works with an application running on your phone. I use SkyFi and SkySafari on my iPhone, iPad or Mac to easily select target objects and slew the telescope to them. Celestron also has a similar device and phone software.

Mount

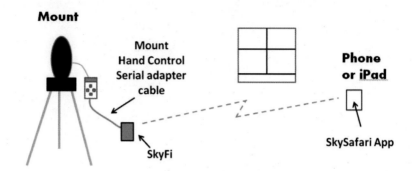

d. Attaching a serial cable to the hand controller and connecting it to a dedicated Wi-Fi device that works with an application running on your computer. I use SkyFi and SkySafari on my Mac to easily select target objects and slew the telescope to them. I have also used SkyFi with a special application COM2TCP I run on the computer that allows the computer to wirelessly connect to the mount and then control the mount with an ASCOM compatible application like Starry Night.

Mount

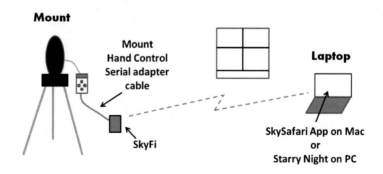

e. Buy a Wi-Fi controlled mount such as one of the new Celestron NexStar Evolution models to control the mount from your phone. This is a good example of how technology is moving forward in ways that will make RVA even more practical!

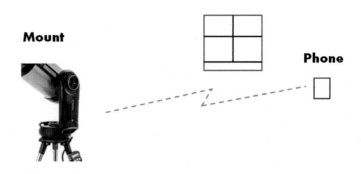

Mount

Phone

f. Use a mount with automatic alignment capability, or add the Celestron StarSense Accessory if you have a compatible Celestron mount so you can perform an automatic alignment without having to be at your telescope. This process supplements the other methods already listed allowing you to set up for remote use before you even turn anything on. When ready, you turn on your system and let the StarSense Accessory perform the alignment while you move back inside and get settled. This is another case of new affordable technology that can make RVA easier.

42. So, is there a way to do RVA with just one cable?

Yes, by setting up a powered hub at your telescope and using the right kind of USB extension cable. I use a Tripp Lite U222-

007-R USB2.0 Powered Hub ($23) at my telescope, and plug into it my camera control cable(s), video capture device and hand controller serial to USB cable. I also have a JMI USB focus control that I sometimes plug in to the powered hub. Using a powered hub makes sure the right USB voltage levels are maintained at the telescope instead of relying on the power from the computer inside. I place the powered hub in a plastic bin to keep all my connections organized. I can either plug the powered hub into AC if available, or include a battery in the plastic bin to power the hub. Once it is powered on, I place the cover back on the bin to keep everything secure.

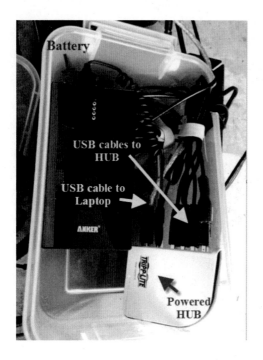

Instead of using the supplied short USB cable for the hub, I use a 35' Tripp Lite U042-036 36 feet High-Speed USB2.0 A/B Active Device Cable ($28) to connect from the USB hub to the computer inside. If I need even further distance, I add a PTC 20 Meters (64 Feet) USB 2.0 Active Extension / Repeater Cable ($35) for a total distance of about 100 feet.

I usually set up my telescope at a fixed place on my driveway (marked with 3 pieces of tape for rough polar alignment) and run the cable into my garage and through a covered hole I made through the wall into my house near the door. I used a short piece of PVC pipe through the wall and attached a cover to the wall over each end using a white plastic weatherproof electrical outlet cover from Lowe's ($5) enabling me to easily open the flap and run the cable through the wall to the inside.

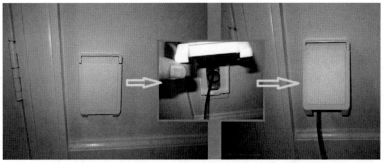

When packing up, after I remove the cable, I stuff a cloth in the PVC pipe and lower the lids. When I set up in my backyard I have a similar arrangement for a permanent access hole through an outside wall but use a metal weatherproof outlet cover ($10) with gaskets on the outside. When not in use these all look like unused outlets.

This setup allows me to switch between the finder camera or main camera view using only one video capture device (see question 20), change my camera settings (see question 40c & 40d) and slew my mount to selected objects using Starry Night (see question 41b) using just one active USB cable from the telescope to inside. I use an HDMI cable sometimes to connect my laptop to my HD TV and use it as an extended monitor to display full screen images from the cameras (see question 45c). This is currently my preferred setup for RVA.

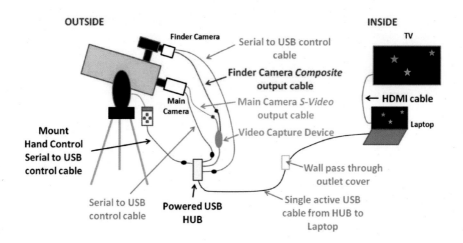

43. So, what are some ways you can do RVA wirelessly?

This depends upon how you plan to use your camera's output. I have found I can use a wireless USB hub to control cameras, the mount and other devices remotely, but trying to send video over a wireless USB hub can be a real problem. Implementing wireless video makes this a more expensive solution than using a USB cable. Here are some basic wireless configurations I have successfully used with technology available at this time. Note that none of these include a wireless remote video finder camera.

1. Wireless USB Hub for computer control, Streaming HDMI for video. I have used the IOGEAR Wireless 4-Port USB Sharing Station GUWIP204 ($75) to connect to my camera's USB control cable, the Hand Control serial to USB cable, and other low bandwidth USB devices to my computer over Wi-Fi. A laptop is used to control the telescope mount (e.g. Starry Night) and adjust the camera settings (e.g. MallinCam control), but not for viewing the video. For Video, I convert the S-Video camera output to HDMI using Portta PETCSHP S-Video to HDMI Converter ($46). If you are using a camera with composite video output you could use the Portta WPETCHP Composite to HDMI Converter ($24) instead. I then steam the HDMI video inside to my TV using the Actiontec MyWirelessTV Wireless HDMI Kit ($180).

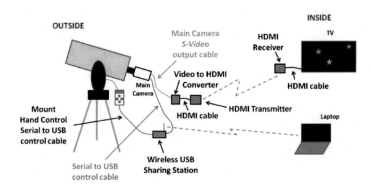

This may seem a little pricey, but I have tried regular video transmitter/receiver devices with disappointing results due to quality and interference issues. (If your camera output is composite video and your distance is under 25 feet, you may be able to use a 2.4GHz composite video transmitter/receiver - see question 15).

2. <u>Wireless mount control, Wireless USB Hub for camera control, Streaming HDMI for video</u>. This is the same setup described in #1 but using SkyFi to control the mount from your Phone or iPad inside over Wi-Fi rather than a computer application like Starry Night. You use the wireless USB Hub for camera settings and use streaming HDMI for wireless video.

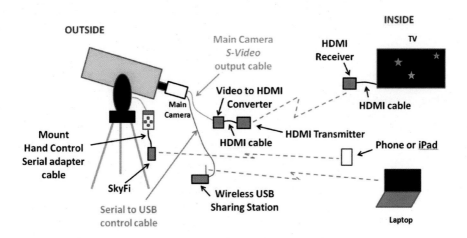

3. <u>Wireless mount control, Wireless camera control, Streaming HDMI for video</u>. This setup will allow wireless viewing, camera control and mount control without a computer (an iPhone or iPad is used to control the mount via SkyFi).

This is the same setup described in #2 but using a camera wireless remote instead of a wireless USB hub. No computer is required with this setup. The MallinCam Jr Pro ($600) is a great Video Astronomy Camera that can be ordered with a wireless remote control allowing you to vary the exposure from 2 seconds up to 99 minutes remotely. I added a wireless remote ($250) to my MallinCam Extreme to be able to operate it wirelessly without a computer.

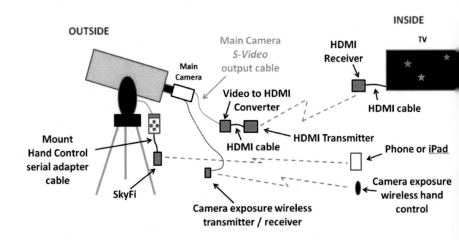

4. Wireless USB Hub for computer control of Mount, DSLR and Video capture. This is the only setup I have found where I can use a wireless USB Hub for everything, and only when using my Canon DSLR. I have used the IOGEAR Wireless 4-Port USB ($75) to connect to my Canon DLSR USB control cable and the Hand Control serial to a USB cable. I use Starry Night to control the mount and Canon software to vary the Canon settings and take pictures. These

picture images are transferred over the USB Wi-Fi without any problems. This is not an automated update RVA setup since you have to initiate each image update manually with the software by taking a picture, and it is storing each image on the hard drive when it is displayed (which you can delete later if you do not want to keep them).

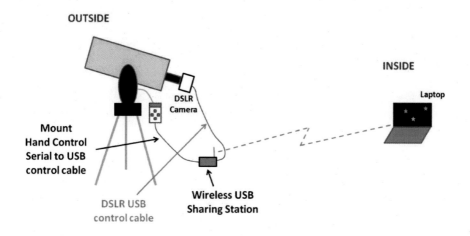

I fully expect advances in technology will provide better and more cost effective solutions for wireless Remote Video Astronomy in the near future

44. So, how do I fix high COM port numbers that are generated when using USB hubs if they cannot be selected in my software application running on a MS Windows system?

You should first test all your USB equipment by directly connecting each one to your computer to get the drivers installed and verify the equipment works. Then plug them into the hub and check that they still work properly.

Devices that do not use a virtual COM port should work fine (e.g. video capture devices). Devices that work as a virtual COM port (e.g. serial to USB adapters) may have a different COM port number when connected to the hub, often as a higher number, requiring you to change the setting for your device's COM port number in the software. Some software applications only support up to a specific COM number. So, what can you do if you connect your device to the hub but the COM port number is out of range for the application? Write down the COM port number you previously used in the application when the device was plugged directly into the computer.

You need to locate the new virtual COM port number for the device on your Windows system when it is plugged into the hub. With the device plugged in to your hub, go to the Control Panel and select Device Manager. Look for the line showing Ports (COM & LPT) devices and click on it to expand it. Note the COM numbers shown (e.g. COM23). Unplug your device and note which entry disappeared. Now plug it back in and when its entry reappears, click on it. Click on the Port Settings tab and then on the Advanced button. The COM Port Number option will show its current COM port number. Click on the pull down list and select a lower COM number. If you wrote down the COM number used when it was plugged directly into the computer, try selecting that COM number (if that number indicated it was In Use, you may have to respond Yes to reuse that number). Or you can pick a free low COM port number. Now unplug your device from the hub, wait for it to be removed from the list, then plug it back in. It should appear as the COM number you changed it to. Restart your application and make sure it is set to use this COM number. From then on the system should "remember" this USB COM port setting as long as you plug into the same USB port on the hub (and the hub's USB cable into the same USB port on your PC).

45. So, how can I view my camera's image on my laptop and TV at the same time?

Here are some methods I have found to use both a laptop screen and TV at the same time:

a. If your camera has two video outputs (like the MallinCam Extreme) you can run two cables inside: one to your computer and one to your TV. You can connect the S-Video to the video capture device at the telescope and the composite video to a long, quality video cable run inside to connect to your TV. If your camera only has composite video out, you can split the signal and run one to the video capture device and one to your TV inside. If you want to run only one cable inside, you could use the second video output (or split the video if you only have one output) as input to a wireless streaming set up directly to your TV as described in question 43.

b. You can also send two video signals over a single Cat5 Ethernet cable using a MuxLab Balum on each end. You can then run the composite signal to your TV and the S-Video to your computer (or vice versa).

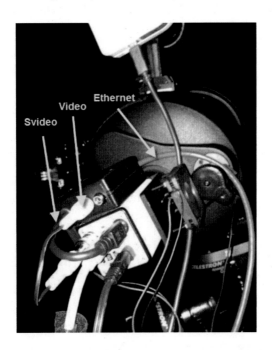

S-Video+Composite Video with Mono Audio Transceiver

S-Video + Composite Video + Audio Balun
Item# 500034

c. If your laptop has HDMI output, another approach for a one cable system is to connect your laptop to your TV's HDMI input and configure your laptop for dual screens with the same image on both. Others can see the same image you are looking at on your laptop.

TV as extended monitor

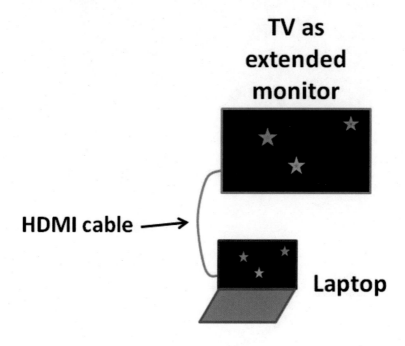

HDMI cable ⟶

Laptop

I have also successfully configured the laptop to use the TV as a separate extended monitor. I use Starry Night on my laptop's screen to select and slew to objects while adjusting and viewing my camera images on the large screen TV. When I have the target object centered and camera exposure and settings adjusted for the best image, I switch the image to full screen which fills up the TV screen.

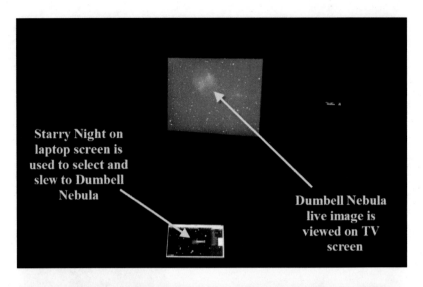

Starry Night on laptop screen is used to select and slew to Dumbell Nebula

Dumbell Nebula live image is viewed on TV screen

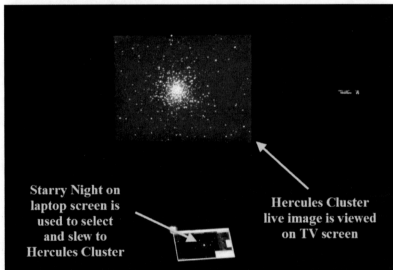

Starry Night on laptop screen is used to select and slew to Hercules Cluster

Hercules Cluster live image is viewed on TV screen

46. So, how do I keep the cables from catching on something while the telescope slews to objects?

This is an important question since you are not at your telescope to watch and keep a cable from getting hung during a slew. I have used different techniques to address this problem. You need to organize the cables that run from the cameras, dew heater, hand control, etc. to power adapters and to the USB hub. I have used Velcro to bundle my cables together and affix them to strategic places on the telescope and mount. This also allows me to reconfigure my cables as needed for different setups.

If you settle on a specific setup, I found using plastic split loom tubing is a way to group cables together, and the slick plastic outer surface slides well with less chance of catching. You can get a 50 foot reel of Install Bay Split Loom Tube 1/2 ($15) that you can cut to various lengths as needed. It is also available in 3/4 inch size.

I make sure the cables from the camera(s) wrap back and attach somewhere near the telescope for strain relief to avoid having any tension on the cables at the camera, which could affect the focus and connections. Depending on the type of mount, I then run the cables over to the arm (Alt/Az) or to the midpoint (GEM) of the mount.

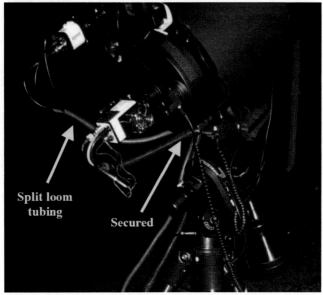

Split loom tubing

Secured

At ground level I split the cables off to the power connections and to the hub USB ports. I place the hub in a plastic container on the ground (or on the wheel mount if you are using wheels) to keep all the connections organized, and run the powered hub's single USB cable out to connect to the controlling computer.

47. So, how do I keep an eye on my telescope and cables from inside when slewing to sky objects?

I have found using an inexpensive Wi-Fi surveillance camera is a good way to keep tabs on your telescope and cables during a slew.

The D-Link Wireless Network Surveillance Camera DCS-932L ($53) is quick and easy to plug in and set on the ground looking up at the telescope. It works with mydlink on the iPhone/iPad or an Android phone. I keep my iPhone/iPad near me and watch it when I slew the telescope to a new target. If I see anything that looks amiss I stop the slew and go take a quick look outside. I rarely have to stop a slew, but being able to see the telescope slew without problems is comforting … and fun! You can also use the D-Link Wireless Pan and Tilt Network Surveillance Camera DCS-5010L ($75) which gives you the ability to pan and tilt the camera as needed. These cameras use your home Wi-Fi, so you will need to have good Wi-Fi signal outside where your telescope will be located to use them.

48. So, how is RVA useful for group viewing?

Video Astronomy is an ideal way to view night sky objects through a telescope with others. It is a great way to involve your family and others in your hobby and is a wonderful teaching aid. You can point out something on a screen displaying the "view" through the telescope that everyone sees together at the same time rather than lining up one at a time to peer through an eyepiece. If you are all outside at the telescope, you can use the techniques for RVA to simplify the operation from a single PC even though you are right next to the telescope. RVA is also great for showing off the night sky to those who are not as inclined to be outside on a cold night (or on a hot humid summer night complete with insects).

There is also an in-between method of RVA that I like to use at times. I set up my telescope for RVA and run the single USB cable into my garage. I put my laptop and a small HD TV on a table in the garage where I can also see the telescope out on the driveway as it slews. I set up the HD TV as an extended monitor as described in question 45c for full screen viewing of objects and use the laptop for control and slewing to objects.

This gives a better protected environment for your laptop and TV so you can avoid dew forming on your laptop and screen, while still being close to the outside where you can simply get up and take a few steps to see the sky where your telescope is pointing. If you go to a viewing site at a remote location, you could set up inside a tent or under a temporary canopy for a similar arrangement. Both of these provide a more protected environment for others while still being able to watch the telescope operate and step out to see the actual sky area of the amazing image they see on the display.

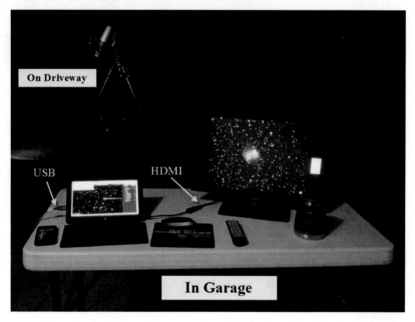

I have also used a combo arrangement where I set up my laptop control area in the garage and set up wireless HD streaming of the image from the telescope to a HD TV inside as described in question 15. This allows some to be close to the outside viewing area while others can remain inside to see images. RVA can even allow your cat to view the Dumbell Nebula - try doing that through an eyepiece!

49. So, are there other things I need to consider to operate remotely inside a vehicle?

RVA allows you to be near your telescope but stay inside your vehicle to stay warm in the winter, avoid bugs in the summer and have refreshments close at hand. You will need to make sure you have the means to operate all your equipment on either the vehicle's battery or a battery pack. See question 51. The single USB cable solution to connect to a laptop inside the vehicle is probably more feasible than a Wi-Fi setup. See questions 41 and 42. You can slightly lower a window to run the USB cable inside (and some ventilation is important, even in the winter).

You might think it would be easy to see your telescope from inside the vehicle as it slews to an object. However, you will have to contend with light reflections on the windows since you have some glow from your laptop. Shining a flashlight through the windshield does not help any…try it and you will see a nice bright reflection right back at you. Duh. Plus, if there are others around, you want to keep use of lights to a minimum. You may be able to see it well enough by using a red light placed near the telescope. Even though you do not need good night vision when using RVA since you are looking at a screen, it helps to dim your LCD screen to its minimum level to reduce its effects on you and others.

Speaking of bright lights, one of your big enemies will be opening a door and the dome lights coming on. You may want to tape your door switch closed if you can during your session to keep this from happening (or temporarily remove the fuse for the dome light).

50. So, what is the difference between adjusting the camera's gain and adjusting the exposure?

The gain setting determines how much amplification is used when generating the output image. Increasing the gain makes faint objects brighter, but also amplifies any noise as well making the overall background lighter. Increasing the camera's exposure time makes faint objects brighter while maintaining a much darker background. For deep sky objects, it is best to have the gain set to off (or a low value) and increase the exposure time to bring out the faint image. If the image starts to blur, or you begin to see star trails due to tracking issues, you may have to increase the gain slightly to reduce the amount of exposure time needed.

51. So, how can I power everything off batteries?

If you do not have AC power available where you plan to set up your equipment for RVA, you will have to provide some means of battery power with the voltage(s) you need. Most of your equipment should run off 12 volt battery power. A portable rechargeable power pack with 12 volt power sockets is a good way to provide this, such as the Celestron Power Tank 17 ($120), which also has both white and red lights to assist during setup and take down.

You can plug your 12 volt power cable for your mount, camera and other 12 volt devices into the Power Tank power sockets (use splitters for additional sockets if needed).

Powered hubs typically run on 5 volts, so you will need a different type of battery that can supply 5 volts for the hub. You can use an Anker Astro E5 15000mAh battery pack ($50) or similar device to provide 5 volts from the battery pack USB charging port. Charge this small power pack before your viewing session and it should power your hub all evening.

You will also need a USB power cable that works with your powered hub. I found a USB cable that came with several charging ends and selected the small barrel connector that matched my hub power connector. The BixPower USB Port Power cable with 3 interchangeable connector tips also worked with my powered hub.

If you use a D-Link Wi-Fi camera to "watch" your telescope from inside, it can be powered off a mini 5V USB battery charger ($20) using a USB power cable such as the StarTech USB to Type M Barrel 5V DC Power Cable ($5).

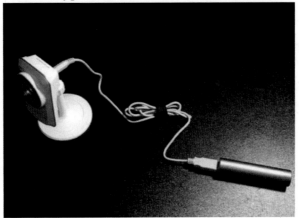

Don't forget that your laptop needs to be fully charged for your evening session as well. If your laptop battery begins to get low and you are in or near a vehicle, you can use an inverter that plugs into your vehicle cigarette lighter socket to provide 115volts for your laptop's AC adapter, which then convert's the power back down to the level needed by your laptop. Be aware that some of the power is lost in these conversions and it puts a drain on your vehicle's battery, so you may only want to do this if your laptop battery gets low.

52. So, should I use my camera or eyepiece when aligning my telescope?

I recommend you initially use an eyepiece to align your telescope as usual and then swap it out for your camera, focus it and move inside. If you leave your focus alone between sessions, you can try performing the alignment using your camera instead of an eyepiece. You will need to have a well aligned finder scope so that when you center the finder on the alignment star, the star will also show up in the field of view of your camera. You can then use fine adjustments to center the alignment star while watching your camera image on your laptop or LCD screen. Using your camera for the alignment process typically produces a more accurate alignment which is undisturbed since you do not have to swap anything after the alignment is finished. I have found it easier to use the camera for the alignment process on my GEM mount since it is already tracking some during the alignment process, while Alt-Az mounts may not start tracking until the alignment process is complete.

53. So, is the Celestron StarSense Auto Align accessory helpful for Remote Video Astronomy?

I have found the StarSense Auto Align Accessory ($330) very helpful for RVA for both Alt-Az mounts and GEM mounts.

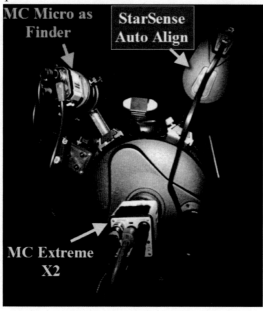

I can install my main camera, connect the cables, power on and let the StarSense perform an alignment while I finish completing my setup and running my cable inside. I can then slew to a bright star, place the focus mask on the end of the telescope to fine tune the focus while looking at an LCD at the scope. When done, I remove the focus mask and proceed inside to begin my session.

This is particularly useful for an Alt-Az mount since there is no need to view the camera image for focusing until after it is aligned. (i.e. Your Alt-Az mount may not start tracking until it is aligned.) At the time of publication, the StarSense did not have the All Star Polar Align function in place yet, but it is planned. I currently use a polar scope to get a good initial manual polar align on my equatorial mount before turning it on and letting StarSense perform an automated alignment. I find the automated alignment also means I no longer find myself getting into odd positions to align my telescope!

54. So, how do you use a polar scope if you leave your equipment on your equatorial mount?

To use a polar scope, you need to position your mount's RA while looking through the polar scope to get the constellations etched inside the polar scope to match their current positions in the sky, and then manually adjust the Alt and Az knobs until Polaris is at the position etched in the polar scope. After that you attach your telescope and other equipment. However, if you keep your RVA equipment in place between sessions, using the polar scope could be problematic. I use a nice application on my iPhone called Scope Help to make this easier.

I first rotate my scope horizontally so I can look through the polar scope in the mount's polar axis.

I then start Scope Help and look at its screen that displays the position of Polaris as it should appear through your polar scope relative to the North Celestial Pole (NCP) at your current

position and time. Note particularly Polaris' angle from the
NCP.

Then I manually adjust the Altitude and Azimuth knobs looking
through the polar scope until Polaris is aligned on the circle at
the angle from NCP shown by the iPhone application. I ignore

the constellations and Polaris etched in the polar scope.

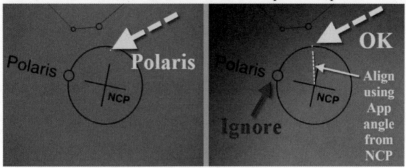

It only takes a short time to check my phone for the correct angle, look through the polar scope and turn the mount's knobs until Polaris is positioned correctly on the circle, then return my mount to its home position. I am now ready to perform an alignment (or a StarSense Auto Align).

55. So, is an All Sky camera useful for Remote Video Astronomy?

An all sky camera can actually be a useful tool when using RVA. I use a Moonglow All Sky Cam ($450) and their All Sky Cam Uploader Kit ($100) that includes a video capture device and software.

It is actually a simple self-contained RVA of its own, albeit a somewhat wide angle view of the entire sky. You can have a monitor on the inside of your house and see a 360 view of the sky above you at any time day or night. Now, you won't be seeing any deep sky objects with an all sky camera, but that is not its purpose. You can see the brightest stars and play back animated archive files to see their motion over time (seconds, minutes, hours and even days). You can see how the constellations rise and set at your site and how nearby obstructions (e.g. trees) affect your views at different times during a night. Through software, you can capture meteors and other events in the night (or day) sky.

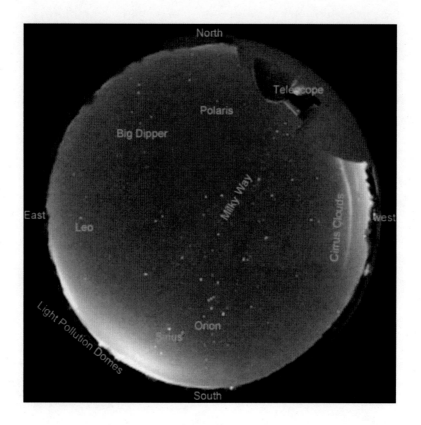

You can check for clouds from inside to help decide if it is even worth setting up outside or not. While you are viewing night sky objects with your RVA system from inside, you can also check your all sky camera view to see if clouds are moving in, or if there are areas of the sky that might be clearing up.

56. So, is there a way I can experience Video Astronomy remotely to see if I want to pursue it myself?

Yes. In fact you can experience Video Astronomy for free in your own home by watching live video images on Night Skies Network. Night Skies Network takes the concept of Remote

Video Astronomy to a different level. You can be a remote observer from anywhere in the world and see someone's real time broadcasts from their telescope using Video Astronomy. Go to http://www.nightskiesnetwork.ca to see if anyone is currently broadcasting and enjoy seeing what they are seeing in real time while listening to their comments. You can even join in the chat discussions associated with what you are seeing. Just log in as a guest and click on a broadcast and enjoy the view with others.

Once you become comfortable with Video Astronomy, you can sign up to become a broadcaster yourself! (My broadcast channel is NightLights). Just go to the website and click on Sign-up to Broadcast. After you select a name and are assigned a channel, click on Site Instructions from your computer that is connected to your Video Astronomy equipment and follow their steps to begin broadcasting your own live viewing session.

Question List

So…

1. Exactly what do you mean by "Beginning Remote Video Astronomy" (RVA)?
2. What is "near real time viewing"?
3. What is the minimum I need to begin Remote Video Astronomy?
4. How do I use this basic setup to get "first light" inside?
5. What are some tips as I get started?
6. What can I do if objects do not appear in the field of view after a GoTo slew?
7. What else would be helpful for an expanded system?
8. Can I adapt webcams for Video Astronomy?
9. Can I adapt security cams for Video Astronomy use?
10. What are the various methods used to display images remotely in Video Astronomy?
11. What are the pros and cons of the different types of cameras used in Video Astronomy?
12. Is there a way to use a finder scope remotely?
13. What may be wrong when using a hand controller extension cable if my mount continues to slew after reaching its target?
14. Is it feasible to do Remote Video Astronomy wirelessly?
15. Can I transmit the video to a receiver inside?
16. How can I control my mount over Wi-Fi?
17. How can I control my camera over Wi-Fi?
18. What is involved in displaying the video on an HD TV?
19. Can I adapt my DSLR for Video Astronomy use?

20. How can I use video cameras for both my main camera and my remote finder from the same PC?
21. What do I need to consider when adjusting the settings on the MallinCam micro when used as a remote finder scope?
22. What should I try out first at the telescope before operating remotely inside?
23. Is there a way to make initial coarse focusing easier?
24. Is there a way to make focusing a video finder scope easier?
25. Would a flip mirror be useful for RVA?
26. If I can't turn the focus knob enough, what do I do?
27. What do you typically use for your RVA system?
28. What does not work very well for Remote Video Astronomy?
29. What are hot pixels, warm pixels and amp glow and what can I do about them?
30. How do you compare using a camera to an eyepiece's magnification and field of view?
31. What are focal reducers and why would I use them?
32. Where should I put focal reducers in the optical train to avoid problems with focus, vignetting, etc.?
33. When should I use a Barlow lens instead of a focal reducer?
34. Does Video Astronomy require an Equatorial mount and/or autoguiding?
35. What is the difference between GoTo accuracy versus tracking accuracy?
36. What is the difference between camera frame integration and frame stacking in software?

37. Can I view a variety of selections from the moon, sun and planets to deep sky objects such as nebulae and galaxies with Video Astronomy?
38. When would I need different thread adapters (T-thread, C-thread, T-Ring, etc.) and nosepieces (1.25" and 2")?
39. What basic filters should I use for Video Astronomy?
40. What are the different ways a camera can be controlled from inside?
41. What are the different ways I can control my mount from inside?
42. Is there a way to do RVA with just one cable?
43. What are some ways you can do RVA wirelessly?
44. How do I fix high COM port numbers that are generated when using USB hubs if they cannot be selected in my software application running on a MS Windows system?
45. How can I view my camera's image on my laptop and TV at the same time?
46. How do I keep the cables from catching on something while the telescope slews to objects?
47. How can I keep an eye on my telescope and cables from inside when slewing to sky objects?
48. How is RVA useful for group viewing?
49. Are there other things I need to consider to operate remotely inside a vehicle?
50. What is the difference between adjusting the camera's gain and adjusting the exposure?
51. How can I power everything off batteries?
52. Should I use my camera or eyepiece when aligning my telescope?

53. Is the Celestron StarSense Auto Align accessory helpful for Remote Video Astronomy?
54. How do you use a polar scope if you leave your equipment on your mount?
55. Is an All Sky camera useful for Remote Video Astronomy?
56. Is there a way I can experience Video Astronomy remotely to see if I want to pursue it myself?

Appendix A
References

- *Deep Sky Video Astronomy* by Steve Massey & Steve Quirk, Springer 2009
- *What is Video Astronomy* at MallinCam.net/what-is-video-astronomy (see also other references at end of the article)
- *Is Video Astronomy the Future of Observing?* by Jim Thompson, Astronomy Technology Today January-February 2013
- *Getting Started with Video Astronomy* at trivalleystargazers.org/talks/2014-02-21.pdf by Curtis V. Macchioni & Nico V. Macchioni, February 2014
- *The Bathinov Focusing Mask* by John Wunderlin, Astronomy Technology Today May-June 2012
- *An Introduction to Astronomical Filters – Filters and Video Astronomy* by Jim Thompson, Astronomy Technology Today November-December 2012
- *Micro-Ex Camera* by Michael Burns and Rock Mallin at www.mallincam.net/uploads/2/6/9/1/26913006/micro_ex_user_manual.pdf
- *Night Skies Network* at NightSkiesNetwork.ca
- Yahoo Video Astro Group
- Yahoo MallinCam Group
- Cloudy Nights Video and Electronically Assisted Astronomy Forum
- MallinCam at mallincam.net
- MallinCam US vendor at mallincamusa.com

Appendix B

MallinCam Micro Crosshair Settings using Privacy Masks

1. Begin by pressing the middle button to display the menu and then use the right arrow to move over to Privacy. Press the middle button to select Privacy.

2. Now press the right arrow until Mask 3 is displayed. Press the down arrow to select Mode and then the right arrow to turn it on. Press the down arrow to select Color then press the right arrow until color 5 is shown. (You may see various shapes displayed as you adjust each mask's values).

3. Press the down arrow until Move is highlighted and then use the right/left arrow to set X to the value shown below. Then use the up/down/left/right arrows to select and set the X values shown for Top, Left, Right and Bottom.

4. With one of the X values highlighted, press the middle button to toggle to the Y values. Then use the up/down/left/right arrows to select and set the Y values shown below for Move, Top, Left, Right and Bottom. You should set the Move value first. Sometimes you have to readjust values until you get all the values set as shown (the settings can interact with one another).

5. When finished with Mask 3, use the up arrow to highlight Mask and then press the right arrow to switch to Mask 4. Repeat steps 2-4 to set the values for Mask 4 as shown below.

6. Repeat the above for Masks 5 and 6.
7. At this point you should see the complete crosshair as a screen overlay.
8. Press the down arrow to highlight Return and then the middle button.
9. Press the right arrow until Exit is highlighted and then the middle button.
10. Use the down arrow to highlight Save/Exit and then press the middle button.

11. You should now see your completed crosshair on the screen.

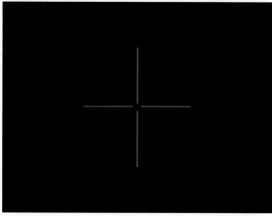

This crosshair will appear whenever you turn on your Micro. You can turn off the crosshairs without losing the settings by going to Privacy and selecting each Mask and turning it off. Then go to Exit and select Save/Exit.

When you want to turn the crosshairs back on, just go back to Privacy and turn on the 4 Masks and then go to Exit and select Save/Exit.

Mask Select	3	4	5	6
Mode	ON	ON	ON	ON
Color	5	5	5	5
Top X	143	97	147	143
Left X	143	97	147	143
Right X	145	140	190	145
Bottom X	145	140	190	145
Move X	144	118	168	144
Top Y	32	65	65	68
Left Y	64	66	66	100
Right Y	32	65	65	68
Bottom Y	64	66	66	100
Move Y	48	65	65	84

Suggestions

Send suggestions, corrections or other feedback to:

BeginSoWTM@aol.com

Made in the USA
Monee, IL
08 January 2022

88474093R00076